Hormone-Disruptive Chemical Contaminants in Food

Issues in Toxicology

Series Editors:
Professor Diana Anderson, *University of Bradford, UK*
Dr Michael D Waters, *Integrated Laboratory Systems, Inc, N Carolina, USA*
Dr Martin F Wilks, *University of Basel, Switzerland*
Dr Timothy C Marrs, *Edentox Associates, Kent, UK*

Titles in the Series:
 1: Hair in Toxicology: An Important Bio-Monitor
 2: Male-mediated Developmental Toxicity
 3: Cytochrome P450: Role in the Metabolism and Toxicity of Drugs and other Xenobiotics
 4: Bile Acids: Toxicology and Bioactivity
 5: The Comet Assay in Toxicology
 6: Silver in Healthcare
 7: *In Silico* Toxicology: Principles and Applications
 8: Environmental Cardiology
 9: Biomarkers and Human Biomonitoring, Volume 1: Ongoing Programs and Exposures
10: Biomarkers and Human Biomonitoring, Volume 2: Selected Biomarkers of Current Interest
11: Hormone-Disruptive Chemical Contaminants in Food

How to obtain future titles on publication:
A standing order plan is available for this series. A standing order will bring delivery of each new volume immediately on publication.

For further information please contact:
Book Sales Department, Royal Society of Chemistry, Thomas Graham House, Science Park, Milton Road, Cambridge, CB4 0WF, UK
Telephone: +44 (0)1223 420066, Fax: +44 (0)1223 420247, Email: books@rsc.org
Visit our website at http://www.rsc.org/Shop/Books/

Hormone-Disruptive Chemical Contaminants in Food

Edited by

Ingemar Pongratz and Linda Vikström Bergander
Department of Biosciences and Nutrition, Karolinska Institute, Huddinge, Sweden

RSCPublishing

Issues in Toxicology No. 11

ISBN: 978-1-84973-189-8
ISSN: 1757-7179

A catalogue record for this book is available from the British Library

Published by The Royal Society of Chemistry,
Thomas Graham House, Science Park, Milton Road,
Cambridge CB4 0WF, UK

Registered Charity Number 207890

For further information see our web site at www.rsc.org

Foreword

Even though effects of both endogenous and exogenous endocrine disruptors
were first observed and described a long time ago, it was not obvious that
these effects were related to disturbances in hormone homeostasis. The observations were not spelled out as disrupters of the endocrine systems. Examples
of such effects are milk production in young boys after consumption of milk
from cows feeding on clover in the spring or the effects observed in birds of
prey and their insufficiency to reproduce successfully. Experimental studies
in the early 1970's induced a prolongation in the estrous cyclicity of mice.
However, over the last two decades endocrine disrupting chemicals and their
effects have become more obvious and are now a matter of major concern.
Numerous national and international reports were produced to address
this issue in the latter part of the 1990's, including one from the UNEP/
WHO 2002.[*]

Thereafter an intense amount of work was devoted to improving our
knowledge and understanding of EDCs and their effects. The present book
"Hormone-disruptive Chemical Contaminants in Food" is one of a few such
attempts to summarize the current state-of-the science. Focusing on EDCs in
food is highly relevant since this is indeed the major source of human exposure
to these chemicals. Over the last decade the EU legislation on chemicals -
REACH (Registration, Evaluation and Authorization of Chemicals) has been
agreed and now implemented. Even though ED effects are not addressed as
such, the endpoints regarding, *e.g.* reproductive toxicity, are highly dependent
on hormone regulations, as well as cancer. In a recent global project related to
REACH and EDCs, the EU have adopted a report on the "State of the Art

---

[*] http://www.who.int/ipcs/publications/new_issues/endocrine_disruptors/en/

Issues in Toxicology No. 11
Hormone-Disruptive Chemical Contaminants in Food
Edited by Ingemar Pongratz and Linda Vikström Bergander
© Royal Society of Chemistry 2012
Published by the Royal Society of Chemistry, www.rsc.org

Assessment of Endocrine Disrupters; 2nd Interim report".[†] The present book is timely and addresses natural and anthropogenic EDCs, verity of endocrine endpoints, hormone systems and novel methodology for studies of EDCs and their effects.

So far slightly more than 140,000 chemicals in commerce have been registered under REACH and by adding a number of natural endocrine active compounds; it is easy to understand the enormous complexity of possible chemical structures interacting. The phrase herein defines, that metabolism is further contributing to the number of potential EDCs; a highly appropriate and relevant conclusion. By looking into the number of endogenous and exogenous compounds plus their metabolites it is clear to see that a very complex exposure scenario exists. Adding to this are two different situations numerous chemicals possessing both persistency and bioaccumulativity characteristics on one hand and chemicals characterized by their pseudopersistency, *e.g.* exposure to polycyclic aromatic hydrocarbons and phthalate esters on the other.

The present book introduces the reader to the extensive complexity of the hormone systems through selected in-depth examples. The important issues of reproduction, fertility-related issues, teratogenicity and cancer in offspring, are addressed and it reaches the conclusion that male reproductive function is at risk. This is very much in agreement with findings of others and supported by both experimental data and observations in wildlife, a mirror of an endpoint that is devastating for populations, independent of species.

It is obvious that food contaminants are linked to sex hormones and their receptors as well as to the aryl hydrocarbon (Ah) receptor. The authors visualize the complexity of compounds interacting with these receptors. A notably high number of anthropogenic chemicals are binding to the Ah receptor. Examples of some very different responses are also given, even though the structural changes in the chemicals exerting these effects are sometimes minimal. Accordingly, it is important to discuss *in silico* methodologies for assessing endocrine effects of chemicals. In view of all the complex results from both *in vivo* and *in vitro* studies it seems appropriate to conclude as herein defined, that *in silico* methodology is a tool that requires interdisciplinary competence for relevant conclusions.

Even though the concept of epigenetics was generally understood some time ago it was not until the last decade that changes in the epigenome were first discussed in relation to ED and their effects in life. The obvious changes in molecules due to methylations seem reasonable causes of changes in hormone system regulations. The examples described indicate a mismatch between programming and real life which is striking.

To study and promote a better understanding of hormone function and the vast number of hormone systems, receptor proteins, transport proteins, and endpoints, it requires the development of new methods for experimental

---

[†] http://www.who.int/ipcs/publications/new_issues/endocrine_disruptors/en/ Authored by: Richard Evans, Andreas Kortenkamp, Olwenn Martin, Rebecca McKinlay, Frances Orton, Erika Rosivatz, 2011

studies. The large number of chemicals, their metabolites and other potential abiotic transformation products requires methods that can be carried out over a short period of time but still give results from fully integrated organisms. Hence financial as well as the ethical aspects need to be considered. Examples of novel developments concerning test systems are given in the present book showing a positive development in this field of ED research.

ED effects in humans and well-established wildlife are a major threat to human health and the sustainable development of wildlife populations. It is therefore urgent to intensify research efforts on EDCs, their effects, mechanisms of action and their synergies with natural and anthropogenic EDCs, in the complex mixtures we are exposed to via food and feed. The issues that need to be solved will require competences from scientists with a deep knowledge in a variety of disciplines. The issue of EDCs also needs to be communicated to policymakers, stakeholders and the public, to alert all of us that management of these chemicals is required. This book *"Hormone-disruptive Chemical Contaminants in Food"* serves, in my view, as a base for communicating the importance of this message of EDCs in food and their *in vivo* interactions.

Åke Bergman
Professor in Environmental Chemistry
Stockholm University

Preface

Chemicals are an integral part of a modern society and consumers benefit from different chemicals on a daily basis. However, there are a number of problems coupled with these products. Because chemicals are present in a wide range of products and goods, consumers are under constant exposure to many of them. In this book, we discuss the health problems that are associated with the presence of chemical contaminants in food. We highlight some of the scientific challenges associated with the characterization of biological effects coupled to exposure to chemical contaminants and some of the needs for future research efforts in this scientific area.

We go through the scientific challenges associated with, for example, detection of chemical contaminants in a complex food matrix, and we discuss some of the problems associated with the current exposure scenarios to chemical contaminants in food, namely the presence of relatively low doses of chemicals with a prolonged exposure time. In this book we have decided to focus our attention on chemicals that somehow have the potential to interfere with the endocrine axis, namely the hormonal pathways that are regulated by transcription factors of different families of proteins. In addition, we also discuss how multidisciplinary scientific approaches are required to develop new knowledge in this area and how new scientific information needs to be "translated" into legislative action in order to develop relevant safety margins of exposure to contaminants in food and food items.

In our opinion, future research efforts in this scientific area face several key challenges, not only of a scientific nature but also other aspects will require considerable attention by all parties in this field. One key aspect is, for example, the need to find new innovative channels of communication between research providers and research users. There is a critical need to shorten the time frame between scientific discovery regarding potential health hazards and legislative implementation.

Issues in Toxicology No. 11
Hormone-Disruptive Chemical Contaminants in Food
Edited by Ingemar Pongratz and Linda Vikström Bergander
© Royal Society of Chemistry 2012
Published by the Royal Society of Chemistry, www.rsc.org

Communication with the larger community is also an area that requires special attention. New ways to share scientific information and risk information need to be developed so that non-experts are able to digest the information that is provided and to make informed decisions regarding risks and benefits coupled with food consumption. These issues represent new challenging and important areas that will require considerable attention in the future.

We would like to thank all the authors that have contributed their expertise that is presented in this book and we hope that the readers will find our views interesting and worthy of further thought.

Linda Bergander and Ingemar Pongratz

Contents

Issues in Toxicology No. 11
Hormone-Disruptive Chemical Contaminants in Food
Edited by Ingemar Pongratz and Linda Vikström Bergander
© Royal Society of Chemistry 2012
Published by the Royal Society of Chemistry, www.rsc.org

Introduction

LINDA VIKSTRÖM BERGANDER AND
INGEMAR PONGRATZ

Department of Biosciences and Nutrition, Karolinska Institute,
SE-141 83 Huddinge, Sweden

1.1 General Introduction

Food consumption is a global issue involving a complex chain of food producers, food handling, transporting and packaging, among others. Today, there is a substantial knowledge of the various hazards ending up in foodstuffs. These hazards range from simple physical hazards to biological hazards, including pathogenic bacteria and naturally occurring toxins, as well as chemical hazards such as pesticides and heavy metals. There is a large heterogeneous group of compounds present, both naturally and man made, in the environment that is causing adverse health effects. These chemical compounds that disturb hormonal pathways are often known as endocrine disrupting chemicals (EDCs).

Exposure to chemical contaminations from the diet is the main critical route for humans, as well as wildlife, to persistent bioaccumulative (fat-soluble compounds with a tendency to build up and reach high levels in an organism) EDCs. Basically, the hormonal or endocrine disruptors are chemicals with the potential to interfere with the function of endocrine systems. Thus, this book will be focusing on diet-derived hazardous substances that disturb/influence nuclear receptor signaling and thereby target the hormonal systems.

Issues in Toxicology No. 11
Hormone-Disruptive Chemical Contaminants in Food
Edited by Ingemar Pongratz and Linda Vikström Bergander
© Royal Society of Chemistry 2012
Published by the Royal Society of Chemistry, www.rsc.org

1.1.1 Endocrine Disruptive Chemicals

Environmental pollutants and their effects on the environment, humans and animals are a significant concern in today's society. During recent years there has been substantial awareness that a variety of environmental pollutants can intervene with the hormonal system. Many man-introduced compounds influence the hormonal system of animals and may be responsible for developmental and reproductive abnormalities seen in wildlife.[1] Natural sources of EDCs are present in various types of foods and are susceptible to metabolic degradation; however, synthetic industrial chemicals, such as inorganic contaminants, agrochemicals, industrial chemicals, plasticizers, plastics, and pharmaceutical agents, that leak into the soil, have the ability to end up in the food chain and thereby bioaccumulate in animals and humans.

The term endocrine disruptor was evolved at the Wingfield meeting in 1991, where a group of researchers with diverse backgrounds was united to discuss the effects of mammalian exposure to environmental chemicals. As a result of the meeting, a consensus statement was set by the participants: "We are certain of the following: a large number of man-made chemicals that have been released into the environment, as well as a few natural ones, have the potential to disrupt the endocrine system of animals, including humans". It was also concluded that the effects of such chemicals are diverse when comparing embryo, fetus, and perinatal organisms to adults and that detectable effects are commonly seen solely in the offspring.[2] Later on, the so-called "endocrine disruptor hypothesis" was published in the book *Our Stolen Future*, which essentially claimed that certain synthetic chemicals interfere with hormone synthesis and, thus, disrupt endocrine networks in animals and humans.[3] To clarify the concept of EDCs, the U.S. Environmental Protection Agency (EPA) defined EDCs as "exogenous agents that interfere with the production, release, transport, metabolism, binding action, or elimination of the natural hormones in the body, responsible for the maintenance of homeostasis reproduction and the regulation of developmental processes".[4]

In 1962, prior to the Wingfield meeting, Rachel Carlson wrote the alarming book *Silent Spring*.[5] This groundbreaking book recapitulates a small fictional world on the road to ruin as a result of accumulated separate disasters; however, all were picked from real life. She discussed the widespread use and the danger of environmental chemicals, such as pesticides and herbicides, on wildlife development and reproduction. This warning for man-made chemicals was first of a kind and, hence, a precursor to the debates on the use of chemical pesticides that later on would result in a ban of the heavily used insecticide DDT, as well as polychlorinated biphenyls (PCBs) in the USA.

1.1.2 Biological Pathways Affected by EDCs

The scientific community has become increasingly concerned that humans experience health problems and wildlife populations are adversely affected following exposure to chemicals that interact with the endocrine system. A well functioning endocrine system, a hormonal balance, is a central function and a key issue for maintaining physiological homeostasis and a healthy body. One hormone in imbalance affects other hormones in the body.

The basics of the endocrine system are a number of glands that secrete the chemical messages that we call hormones. The major glands of the endocrine system are the hypothalamus, pituitary, thyroid, parathyroid, adrenal, pineal body, and the reproductive organs (ovaries and testes). These glands release a diversity of hormones directly into the bloodstream, where they target an organ and thereby regulate various processes, like growth, metabolism, development, reproduction, and sexual characteristics.

Hormones exert their action through a range of receptors by a lock-and-key model. These receptors are either (i) membrane bound and linked to ion channels, G-proteins, or enzymes or (ii) intracellular and localized in the nucleus or the cytosol. The membrane-bound receptors mediate the cellular response to hormones either by a depolarization of the membrane or by the generation of so-called second messengers or signal transducing molecules. The intracellular receptors, on the other hand, mediate the cellular response by modulating gene expression in target cells.

Absolute receptor specificity is rarely encountered and, hence, hormone receptors may bind exogenous compounds other than their primary endogenous ligands. By this means, a fraction of exogenous pharmaceuticals, as well as agricultural/industrial chemicals or EDCs released into the environment, may bind to hormone receptors and activate the receptor in a manner similar to endogenous compounds. These compounds may also interfere with the binding and actions of endogenous ligands without activating the receptor itself.[6]

The basis for endocrine disruption is not fully known; however, knowledge of the mechanism of action of the hormonal-disrupting chemicals is advancing. Basically, it has developed from a narrow hormone receptor point-of-view to a broader approach of targets related to nuclear receptors, non-nuclear steroid hormone receptors, non-steroid receptors, and orphan receptors, as well as enzymatic pathways concerning steroid metabolism. With the wide range of EDC actions, it is hard to distinguish if the diversity of the biological end points induced by EDCs is due to direct or indirect effects of EDC exposure. However, it has been noticed that a majority of these compounds work by mimicking or interfering with the normal actions of endocrine hormones, including estrogens, androgens, thyroid, hypothalamic, and pituitary hormones. Today, chemicals that mimic or antagonize the female estrogenic hormones, the male androgenic hormones, or the thyroid hormones are gaining the most attention.

1.2 Chemicals Contaminating Our Food

There is a broad spectrum of compounds with a wide range of physical properties compromising endocrine disrupting qualities. Owing to the heterogeneity of EDCs, the only similarity being small molecular masses, it is hard to predict possible endocrine disrupting actions of chemicals.[7] Also, as only a fraction of all potentially physiological disrupting compounds in the environment has been investigated, it is challenging to develop techniques to improve the analysis of these compounds.[8] However, the European Union has summarized a report categorizing chemicals on the basis of available evidence of endocrine disrupting effects. Out of 146 high production volume chemicals and/or highly persistent substances, a group of 60 compounds were considered to have high exposure risks regarding endocrine disruption.[9]

As previously described, a large diverse group of hormone-disruptive chemicals is present naturally, as well as man introduced, in the environment. The natural source of EDCs present in various types of foods has been termed phytoestrogen and defined as any plant compound structurally and/or functionally similar to ovarian and placental estrogen and its active metabolites.[10] Apart from the beneficial health effects of phytoestrogens, including the prevention of cancer, atherosclerosis, menopausal syndromes, and bone density loss, adverse health effects of phytoestrogens are emerging with a potential for endocrine disruption.[11]

The phytoestrogens are divided into two major classes: the polyphenolic flavonoids and the lignans. Among all, the most well-known phytoestrogens are the soy and chickpea isoflavones genestein and daidzein and the clover-derived comuestrol, as well as the lignans, mainly found in grains, seeds, and other fiber-rich foods. Flavonoids are highly consumed by the Asian population whereas the lignans are generally more consumed by Europeans.[12] Both flavonoids and lignans are, however, in general ingested as precursors and converted into active compounds by the microbial system.[13]

As described above, the synthetic group of EDCs is widespread in the ecosystem and a variety of these chemicals has been designed to be long lasting in the environment and are, therefore, not easily degraded. Chemicals that were banned a long time ago are still found in the ecosystem, even at locations far away from where they were initially utilized.[14] Typical synthetic contaminants found in the food chain are industrial chemicals, like combustion by-products including PCBs and dioxins, the polybrominated flame-retardants (PBB and PBDEs), and biocides and pesticides including tributyltin and DDT. Other contaminants in food are inorganic compounds like heavy metals and metalloids, such as mercury, cadmium, lead, and arsenic, as well as pharmaceutical or synthetic hormones, such as diethylstilbestrol (DES). A more recently emerging food-contaminating group is the chemicals originating from packing materials. Phthalates, used as plasticizers, and the plastic monomer bisphenol A (BPA), a high production chemical used all over the world, are leaking out from packing materials, subsequently resulting in animal and human exposure.[15] Other emerging chemical contaminants are the ubiquitously used polyfluorinated chemicals (PFCs), like perfluorooctanesulfonic acid (PFOS), in non-stick coatings and food packing. Recent reports are showing bioaccumulation in wildlife and humans of PFOS, as well as endocrine disrupting properties.[16,17]

1.2.1 Sources and Routes of Food Contaminating Chemicals

Over the past century, humans have introduced substantial amounts of chemical substances into the environment, all with an unpleasant ability to enter the body by absorption. The exposure route proceeds by means of inhalation, *i.e.* absorption through the lungs, absorption through the skin and, most importantly, oral ingestion and absorption through the digestive system. Chemical hazards can be found in the natural environment (air, water, soil) from industrial and environmental pollution. They also exist in food products as natural chemicals or chemicals produced during manufacturing and processing procedures and eventually arise in the food supply. All of these chemicals have a risk to affect health adversely. Fetuses, children, and adults are all at high risk of exposure to chemicals originating from contaminated food absorbed in the digestive system and ending up in blood and stored in tissues.

High exposure of persistent EDCs is associated with high consumption of fatty food; consequently a reduction of dietary fat should correspond to a reduction of organic contaminants.[18] Chemicals are preferentially eliminated from the body by making them water soluble and ready for excretion. Some chemicals, the hydrophobic or fat soluble, prefer fatty surroundings and accumulate in tissues rather than being extracted as a water-soluble product. This means that EDCs have the ability to accumulate and thereby concentrate in tissues and that a low level of chemicals in water, soil, or plants can be concentrated higher up in the food chain owing to elevated consumption. One of the major routes of removing accumulated chemicals is through breast milk to a nursing baby. According to a recent study, there is a country-specific pattern of EDCs in breast milk, where the Danish population has a higher exposure to persistent bioaccumulative chemicals than, for example, Finnish mothers. These results are interestingly correlated with a higher frequency of male endocrine disrupting disorders in Denmark.[19]

In 2006, the WWF launched a report about the food link in the chain of contamination.[20] A wide range of food items, like dairy products, meat, fish, bread, honey, and olive oil, selected from seven EU countries, was analyzed and toxic residues were found in all products. High amounts of phthalates were detected in olive oil, PBDEs in minced beef, and DDE, PFOS, and PCBs in pickled herring.

Mercury has been well known as an environmental pollutant for several decades. It is a global pollutant of major concern with numerous environmental sources, such as the mining and charcoal industries as well as the healthcare sector. Mercury enters the food chain as a more toxic form, methylmercury (MeHg). It is converted from elemental mercury released into the environment by bacteria and bioaccumulates, especially in fish.[21] Exposure to MeHg in vertebrates results in, among all, embryo toxicity, endocrine disruption, and altered reproductive behavior.[22] In the early 1970s a major MeHg poisoning catastrophe occurred in Iraq, owing to the use of MeHg as a fungicide for treatment of seed grain.[23] Interestingly, a recent publication described mercury as a compound that radically changes birds' mating behavior and a cause of homosexuality in male ibises.[24]

The naturally occurring EDCs, or the phytoestrogens, described above are bioactive compounds structurally and/or functionally similar to the endogenous estrogen and its active metabolites and, thus, have hormone-like activity. Phytoestrogens usually show a weaker estrogenic activity than the endogenous hormones but are highly consumed, especially soybeans, nuts, and seeds.[12] Hence, there is an emerging concern regarding the daily exposure to soy infant formulas, which may result in exposure of infants to high amounts of isoflavones.[25]

1.2.2   Vital Topics Regarding Dose and Mixture Effects

A lot of evidence has been presented during the years regarding the large number of chemicals that pose a risk to human and environmental health. However, the full range of EDCs present in the environment remains largely unknown. The knowledge on EDCs is complicated when considering the effects of chronic low-dose exposure of chemicals through the diet. Extra care should be taken when addressing the exposure during early life, as fetus, infants, and young children, since chronic low-concentration exposure to EDCs is often seen in long-term health effects. In that way, as reproductive disorders typically affect younger individuals, the implications may only become apparent many years later. The concern for chronic low exposure of EDCs is also amplified when considering exposure to a mixture of compounds. We are all exposed, daily, to a cocktail of chemicals, although the knowledge of such combinational exposure is still under elucidation and the regulations of chemicals in food are based on tests of the individual compounds. However, there is growing concern that the substances in combination may cause a greater risk compared to exposure to individual substances. Toxic effects may occur during simultaneous exposure of chemicals owing to chemical interactions that alter the absorption, biotransformation, or excretion of one or both of the interacting chemicals. Up to now, risk assessments for toxicity are set on separate chemicals based on NOAEL, "no adverse effect level". This means the highest dose at which no adverse effects have been detected; however, combinational effects can be expected even at doses well below NOAELs, provided a sufficiently large number of chemicals is present.[26] Different chemicals in a mixture could affect each other in either an antagonistic (weakening) way or in a synergistic way, *i.e.* the combined effects are stronger than the additive effect given the knowledge of each chemical's toxic quality alone.

Some chemicals are produced in large volumes and one of those receiving much attention today is the monomer and component of polycarbonate plastics and plasticizers, bisphenol A (BPA). It is one of the highest volume chemicals produced worldwide, with widespread human exposure. Several "low-dose" studies have suggested that exposure to BPA in the period immediately before and after birth is associated with a selection of abnormalities in the female reproductive tissues.[27]

1.3 Adverse Effects of EDCs

Already in 1960 there were reports describing a food contamination incident occurring down the food chain. Plankton in the Clear Lake north of San Francisco was assimilating DDD, the "less harmful" successor of DDT, and the poison was concentrated up the food chain and transferred on to the larger animals, ending up with a dying population of grebe.[5] The heavily utilized insecticide DDT was initially extensively introduced into the environment in the 1940s. It was among the first chemicals reported to adversely affect endocrine functions and embryonic survival of bald eagles as result of eggshell thinning.[28] In 1966, the Swedish chemist Jensen defined PCB as a new chemical hazard[29] and later on published a report on PCBs and DDT in marine animals of Swedish waters, another example of bioaccumulation up the food chain. Both reproduction and immune functions were disturbed by PCBs in the food chain in Baltic seals (reviewed in Vos *et al.*[1]). Additionally, alligators living in Lake Apoka, USA, suffered from distorted sex organ development and function due to a major pesticide spill.[30]

Recently, 18 years after the Wingfield statement of the potential of environmental chemicals to disrupt the endocrine system, the Endocrine Society published an updated statement of the posed threat of EDCs on human health. In this, the authors "present evidence that endocrine disruptors have effects on male and female reproduction, breast development and cancer, prostate cancer, neuroendocrinology, thyroid, metabolism and obesity, and cardiovascular endocrinology" and they implicated EDCs "as a significant concern to public health".[7]

1.3.1 Vulnerable Populations at Risk

As mention above, there is a concern for chronic low-dose exposure and mixture exposure of EDCs that includes both wildlife and humans. However, it is alarming when it comes to human exposure occurring during critical periods of development since it can cause irreversible effects on, for instance, sex organ development and reproductive behavior. Moreover, sensitivity to EDCs might also be affected by genetic factors that determine specific metabolic pathways, as well as lifestyle factors such as dietary habits.

Some groups of the population are more sensitive to exposure to EDCs than others, like fetuses, children, and pregnant women. The fetus is particularly sensitive to changes in hormone levels, owing to the ongoing development of organs and neural system. The fetus is susceptible to EDCs transferred across the placenta from the blood of an exposed pregnant mother, with an increased risk of birth defects. The sensitive time of exposure for the fetus expands until a few weeks after birth, the postnatal period. During this stage, the baby is susceptible to exposure of bioaccumulated EDCs through the breast milk of the mother. Therefore, small amounts of endocrine-disrupting chemicals that might not affect the mother may still be harmful to the baby. Growing children that are still under development are also a risk group for exposure to EDCs.

Basically, the timing of exposure and the mixture of exposure compounds are of particular concern regarding potential health affects from EDCs. Interestingly, in an ongoing study in the U.S., approximately 100 000 children are planned to be examined from birth to age 21 to study the long-term health effects due to environmental and genetic factors.[31]

1.3.2 Biological Pathways Targeted by EDCs

During two decades in the mid-20th century the hormonally active synthetic estrogen DES was prescribed to pregnant mothers to prevent miscarriage. However, it was soon discovered that the effects of the medication were devastating. Many of the children exposed to DES before birth, both girls and boys, suffered from reproductive and immune system disorders and these problems came to light when several cases of a rare form of vaginal cancer were reported in young girls in the early 1970s.[32] This incident has served as a backbone for the fact that hormone-like substances can adversely affect humans, where DES serves as an endocrine disruptor. Hormone-disrupting compounds have been linked to reproductive and developmental disorders in several species, *e.g.* mammals, birds, reptiles, and invertebrates. Abnormalities, caused by chemicals, in wildlife and laboratory animals comprise a weakening of several populations, altered immune function, disturbed neurological development, decreased fertility, altered reproductive organs, demasculinization and feminization, behavior changes, disturbed thyroid function, and tumor development, among others (reviewed in Vos *et al.*[1] and Hamlin and Guillette[33]).

As pointed out by Guillette and Guillette,[30] several abnormalities seen in the reproductive system of various species of wildlife correlate with similar abnormalities of rising incidences in human populations, although the relation between EDC exposure and human health effects are not yet clearly elucidated.

There have been several reports over the past decade describing disorders in male reproductive health, suggested to be caused by environmental factors. For example, development abnormalities of the male reproductive tract, like undescended or maldescended testis (cryptorchidism), defects of the urethra (hypospadias), problems with semen quality and sperm count, and testicular cancer, are all linked as testicular dysgenesis syndrome (TDS).[34] However, there are limited data concerning women's reproductive health, even though the conception rates in women have declined by 44% in the USA since the 1960s, as emphasized by Hamilton.[35] Proposed female reproductive disorders include early pubertal development, polycystic ovary syndrome, spontaneous abortions, breast cancer, reduced fertility, and endometriosis (spreading of cells from inside the uterus) (reviewed in Diamanti-Kandarakis *et al.*[36]). It has also been suggested that EDCs may be causing altered sex ratios, with fewer males born in humans.[37]

An additional EDC target is the thyroid neuroendocrine system, which is essential for normal brain and proper physiological development.[38] Also, a new emerging area concerning EDCs and human health is the connection to metabolic syndrome, type 2 diabetes, and obesity. The obesity rates have been increasing considerably during the past three decades in both adults and children.[17] Different EDCs have been shown to stimulate fat cell (adipocyte) differentiation, which could lead to the acceleration of lipid uptake.[39]

As discussed in a Chapter 4, it has also been recognized that EDCs have the ability to cause transgenerational effects, meaning that these chemicals not only influence the directly exposed individuals but also affect future generations.

1.4 Challenges in Food Safety

Food and food consumption have evolved from a national or regional commodity to a truly global product and an international plate that can include products from different regions around the world. Furthermore, food is a general product. We all eat food and, regardless of social position, the food we purchase and consume is a potential source of health and disease. The food trade has also changed considerably and has evolved from a "local" product to a global product. The producer is far from the consumer and food is transported across large distances. To be able to deliver the products to consumers they have to be packaged, preserved, and treated so that it is maintained in a state suitable for consumption. This treatment may result in the intentional or unintentional presence of chemicals in food.

1.4.1 What is the Current Risk?

As described previously, chemical exposure poses a risk for humans. Chemicals are an integral part of modern societies and, in fact, some estimates suggest that currently over 70 000 different chemical are marketed in the European Union today. The major problem, however, is that there is very limited scientific information regarding the effects on human health that exposure to these chemicals may impose and that the unwanted presence of chemicals in food represents a possible health problem.

A number of different studies have demonstrated that industrial chemicals are indeed present in humans. However, at the same time it is important to state that in the vast majority of studies the concentrations of these chemicals are low. Food quality has in general improved regarding, for example, bacterial contamination. In addition, most chemical contaminants and additives are present in relatively low levels in food. At this low exposure level, rapid toxic effects are rare. However, even at these low doses, chemical contaminants do pose a considerable risk to humans and there are currently considerable gaps in scientific information regarding the effects of chemical contaminants in food.

The presence of endocrine disruptors is a source of general concern. Currently, food regulators do not include endocrine disruption as a biological endpoint, which in fact means that interference with hormonal signaling is not included when the risk that chemical contaminants pose to humans is assessed.

In addition, exposure to chemical contaminants in food leads to continuous exposure that may last for decades. This implies that interference in the hormonal system is constant and can also be a source for human disease.

One of the key problems facing the scientific community, legislators, and food producers is the cocktail effect, described in Section 1.2.2. Also, as will be discussed in Chapter 12, certain human populations display an elevated sensitivity to the endocrine disrupting actions of chemical contaminants. The problem is that we currently lack scientific methodology to assess the effects of compounds in combination and thus the risk that compounds pose on human health can be severely underestimated.

Clearly, in the future, new legislative paradigms are needed to protect consumers. Today's threshold values used to limit exposure to chemicals are mostly derived from toxicological experiments. These experiments are based on short-term exposure to individual contaminants at relatively high doses. The current exposure scenario, however, displays a different pattern, with low levels of exposure over prolonged time periods.

In addition, it is important to introduce endocrine disruption as a legislative parameter. Countless scientific studies have firmly linked hormonal imbalance with increased risk for disease development. Moreover, the wide range of diseases influenced by hormones should make the need for new and modern legislation a top priority.

1.4.2 Animal Studies and *In Vitro* Studies

Currently, risk assessment serves as the foundation for legislative action. The maximal dose that food items may contain of a given chemical is developed from risk assessment studies.

Risk assessment is in turn primarily derived from animal experiments. Typically, animals (mainly rodents) are treated with chemicals at different doses and for different time periods and the biological effects of this exposure are characterized. However, this approach is expensive and there are clear ethical considerations that need to be taken. However, a major concern is time. Animal experiments are time consuming and, given the large numbers of chemical that are currently in the market, it will take a very long time until all relevant chemicals are tested.

Clearly, it is important to find a new scientific approach to protect the health of consumers. One possibility is to introduce scientific data that are not derived from animal experimentation into the formal process. Another possibility is to introduce *in silico* and *in vitro* data into consideration, both in the risk assessment and in the legislative process. Using *in silico* information it is possible to predict that chemical contaminants may bind to intracellular receptors and proteins and modulate or disturb, for example, hormonal signaling pathways. In addition, using *in vitro* approaches such as established cell-lines and other non-animal models it is possible to identify those cellular pathways that are disturbed by chemical contaminants in food. However, these experiments do not provide information regarding the actual biological endpoint that is affected by a given contaminant. This lack of clear connection to biological endpoints has limited the use of *in vitro* and *in silico* data in the legislative process to date.

However, with the development of new and powerful methodologies, such as whole genome approaches, and the development of novel scientific information demonstrating the involvement of the hormonal system, it is important to develop new approaches in order to be able to introduce these scientific data into guidelines.

1.4.3 Future Challenges in Food Safety

Consumer health protection is a key aspect in the field of food production and consumption. The presence of large numbers and amounts of chemicals in western societies imposes new challenges to the regulatory and scientific communities that have currently not been solved to a satisfactory degree. There is a need to develop new scientific models to study health implications regarding exposure to chemical contaminants in food.

There are, in fact, several challenges essential to be addressed in the near future. A key aspect is to develop new risk assessment methods that can take into account the mixture effects compared with today's methodology that assesses the risk of individual compounds. In addition, it is necessary to develop new, fast, and sensitive methods for detection. In contrast to currently available methods, new methods should provide information regarding the effects of chemical contaminants on hormonal signaling pathways and should also incorporate the scientific knowledge that has been developed in other scientific fields (such as medicine) regarding the mechanisms of action of chemical contaminants.

To meet these challenges, new scientific information needs to be generated and, more importantly, new strategies need to be developed to protect consumer health. These new strategies also need to accommodate the challenges of the current exposure scenario, namely low dose and prolonged exposure of EDCs.

Today there are clear limitations on the number of chemicals that the scientific community can test using the classical toxicological animal-based approach. In fact these limitations are purely practical. The sheer numbers of chemicals that are present in the market today compared to the practical capacity to perform classical animal experiments implies that new testing strategies need to be developed as soon as possible. In addition, we expect that, in the future, different scientific disciplines will need to collaborate to implement new scientific findings in the field of food safety. In addition, the communication between the scientific community and different stakeholders needs to receive special attention.

In fact, new communication paradigms need to be developed to facilitate the information transfer between the scientific community and industry and/or regulatory agencies. Recent scientific information shows that chemical contaminants can interfere with hormonal signaling pathways. This interference can in turn lead to disease development, which in many cases will not manifest itself directly after exposure. Instead, interference with the hormonal signaling pathways gives rise to increased risk to develop diseases that may manifests themselves later in life. In fact, there is ample evidence which suggests that many lifestyles diseases such as obesity, allergy, and other common western diseases may be a result of hormonal interference.

The problem, however, is that it is currently difficult to regulate "risk for disease". Thus, it is important to develop new scientific knowledge using a multidisciplinary approach to protect consumer health from debilitating diseases.

References

1. J. G. Vos, E. Dybing, H. A. Greim, O. Ladefoged, C. Lambre, J. V. Tarazona, I. Brandt and A. D. Vethaak, *Crit. Rev. Toxicol.*, 2000, **30**, 71–133.
2. C. C. Colborn T, eds. *Chemically induced alterations in sexual and functional development: the human/wildlife connection*, Princeton Scientific Publishing, Princeton, NJ, 1992.
3. T. Colborn, D. Dumanoski and J. P. Myers, *Our stolen future*, Plume/Penguin Books, New York, 1996.
4. R. J. Kavlock, G. P. Daston, C. DeRosa, P. Fenner-Crisp, L. E. Gray, S. Kaattari, G. Lucier, M. Luster, M. J. Mac, C. Maczka, R. Miller, J. Moore, R. Rolland, G. Scott, D. M. Sheehan, T. Sinks and H. A. Tilson, *Environ. Health Perspect.*, 1996, **104**(Suppl 4), 715–740.
5. R. Carson, *Silent Spring*, Houghton Mifflin Co, Boston, 1962.
6. A. le Maire, W. Bourguet and P. Balaguer, *Cell Mol. Life Sci.*, **67**, 1219–1237.
7. E. Diamanti-Kandarakis, J. P. Bourguignon, L. C. Giudice, R. Hauser, G. S. Prins, A. M. Soto, R. T. Zoeller and A. C. Gore, *Endocr. Rev.*, 2009, **30**, 293–342.
8. R. A. Trenholm, B. J. Vanderford, J. C. Holady, D. J. Rexing and S. A. Snyder, *Chemosphere*, 2006, **65**, 1990–1998.
9. D. E. EUROPEAN COMMISSION, *Towards the establishment of a priority list of substances for further evaluation of their role in endocrine disruption*, Delft, 2000.
10. P. L. Whitten and H. B. Patisaul, *Environ. Health Perspect.*, 2001, **109**(Suppl 1), 5–20.
11. P. Moutsatsou, *Hormones (Athens)*, 2007, **6**, 173–193.
12. Z. H. Liu, Y. Kanjo and S. Mizutani, *Water Res.*, **44**, 567–577.
13. R. A. Dixon, *Annu. Rev. Plant Biol.*, 2004, **55**, 225–261.
14. W. L. Lockhart, R. Wagemann, B. Tracey, D. Sutherland and D. J. Thomas, *Sci. Total Environ.*, 1992, **122**, 165–245.
15. J. Muncke, *Sci. Total Environ.*, 2009, **407**, 4549–4559.
16. A. A. Jensen and H. Leffers, *Int. J. Androl.*, 2008, **31**, 161–169.
17. C. Casals-Casas and B. Desvergne, *Annu. Rev. Physiol.*, **73**, 135–162.
18. J. L. Domingo, *Crit. Rev. Food Sci. Nutr.*, **51**, 29–37.

19. K. Krysiak-Baltyn, J. Toppari, N. E. Skakkebaek, T. S. Jensen, H. E. Virtanen, K. W. Schramm, H. Shen, T. Vartiainen, H. Kiviranta, O. Taboureau, S. Brunak and K. M. Main, *Int. J. Androl.*, **33**, 270–278.
20. WWF, Editon edn., 2006.
21. J. T. Trevors, *J. Basic Microbiol.*, 1986, **26**, 499–504.
22. S. W. Tan, J. C. Meiller and K. R. Mahaffey, *Crit. Rev. Toxicol.*, 2009, **39**, 228–269.
23. F. Bakir, S. F. Damluji, L. Amin-Zaki, M. Murtadha, A. Khalidi, N. Y. al-Rawi, S. Tikriti, H. I. Dahahir, T. W. Clarkson, J. C. Smith and R. A. Doherty, *Science*, 1973, **181**, 230–241.
24. P. Frederick and N. Jayasena, *Proc. Biol. Sci.*, 2011, **278**, 1851–1857.
25. H. B. Patisaul and W. Jefferson, *Front. Neuroendocrinol.*, **31**, 400–419.
26. A. Kortenkamp, *Int. J. Androl.*, 2008, **31**, 233–240.
27. R. R. Newbold, W. N. Jefferson and E. Padilla-Banks, *Environ. Health Perspect.*, 2009, **117**, 879–885.
28. R. A. Faber and J. J. Hickey, *Pestic. Monit. J.*, 1973, **7**, 27–36.
29. S. Jensen, *New Scientist*, 1966, **32**, 612.
30. L. J. Guillette, Jr. and E. A. Guillette, *Toxicol. Ind. Health*, 1996, **12**, 537–550.
31. N. R. C. Panel To Review The National Children's Study Research Plan, Institute Of Medicine, *The National Children's Study Research Plan: A Review*, National Academies Press (US), Washington (DC), 2008.
32. A. L. Herbst, H. Ulfelder and D. C. Poskanzer, *N. Engl. J. Med.*, 1971, **284**, 878–881.
33. H. J. Hamlin and L. J. Guillette, Jr., *Syst. Biol. Reprod. Med.*, **56**, 113–121.
34. J. Toppari, H. E. Virtanen, K. M. Main and N. E. Skakkebaek, *Birth Defects Res. A: Clin. Mol. Teratol.*, **88**, 910–919.
35. B. E. Hamilton and S. J. Ventura, *Int. J. Androl.*, 2006, **29**, 34–45.
36. E. Diamanti-Kandarakis, E. Palioura, S. A. Kandarakis and M. Koutsilieris, *Horm. Metab. Res.*, **42**, 543–552.
37. I. del Rio Gomez, T. Marshall, P. Tsai, Y. S. Shao and Y. L. Guo, *Lancet*, 2002, **360**, 143–144.
38. T. R. Zoeller, *Hormones (Athens)*, **9**, 28–40.
39. M. A. Elobeid and D. B. Allison, *Curr. Opin. Endocrinol. Diabetes Obes.*, 2008, **15**, 403–408.

CHAPTER 2

Persistent Organic Pollutant Levels in Commercial Baby Foods and Estimation of Infants Dietary Exposure

KARL-WERNER SCHRAMM[a,b] AND MARCHELA PANDELOVA[a]

[a] Helmholtz Zentrum München, German Research Center for Environmental Health, Institute of Ecological Chemistry, Ingolstädter Landstrasse 1, 85764 Neuherberg, Germany; [b] Technical University of Munich, Center of Life and Food Sciences Weihenstephan, Research Departments Biosciences, Weihenstephaner Steig 23, 85350 Freising, Germany

2.1 An Historical Perspective: Persistent Organic Pollutants and Children

Throughout the 20th century, and especially after the Second World War, the chemical industry expanded rapidly and in the 1960s it became clear that certain chemicals had become widely dispersed in the environment. Some are considered to be toxic pollutants and although most of these chemicals have been banned in almost all industrial countries since the late 1980s and are therefore no larger produced, they are still used to a limited extent, both legally and illegally. Today, persistent organic pollutants (POPs) are detected ubiquitously around the globe, even in areas far from any human habitation.

Issues in Toxicology No. 11
Hormone-Disruptive Chemical Contaminants in Food
Edited by Ingemar Pongratz and Linda Vikström Bergander
© Royal Society of Chemistry 2012
Published by the Royal Society of Chemistry, www.rsc.org

In response, the Stockholm Convention, adopted in 2001 and brought into force in 2004, requires signatory parties to take measures designed to eliminate or at least significantly reduce the release of POPs into the general environment.[1] The 12 POPs recognized as causing adverse effects on humans and the ecosystem can be divided into three categories: (1) pesticides, including aldrin, chlordane, DDT, dieldrin, endrin, heptachlor, hexachlorobenzene, mirex, and toxaphenes; (2) industrial chemicals, such as hexachlorobenzene and polychlorinated biphenyls (PCBs); and (3) byproducts like hexachlorobenzene, polychlorinated dibenzo-*p*-dioxins (PCDDs), polychlorinated dibenzofurans (PCDFs), and PCBs (www.chm.pops.int). Moreover, in May 2009 the POP Review Committee and Conference of the Parties decided to add nine new chemicals to this list, *i.e.* (1) the pesticides chloredecone and lindane; (2) the industrial chemicals hexabromobiphenyl, commercial pentabromodiphenyl ether, commercial octabromodiphenyl ether, perfluorooctanesulfonic acid (PFOS) and its salts, and perfluorooctanesulfonyl fluoride (PFOS-F); and (3) the byproducts α-hexachlorocyclohexane and β-hexachlorocyclohexane.

These persistent pollutants are not only toxic and stable, they may also bioaccumulate. As their degree of chlorination increases, the POPs become more and more hydrophobic; furthermore, even at low levels these compounds can disrupt endocrine functions.[2] Substances such as PCDD/Fs and dioxin-like PCBs (dl-PCBs), to which we are exposed *via* dietary animal fats,[3] can apparently interact with several nuclear receptors.

One of the most important tasks of research concerned with persistent pollutants is to identify the doses which are potentially harmful. At the end of the 1980s a Nordic group reported that, at daily doses below approximately 1 ng per kg body weight, tetrachlorodibenzo-*p*-dioxin (TCDD) does not increase the risk that rats will develop liver cancer appreciably,[4] but higher doses elevate this risk significantly. To convert this no-observed-effect level to a tolerable intake for human beings, these experts employed a safety factor based on the assumption that the average individual could be as much as 10-fold more susceptible to this carcinogenicity of dioxins than the laboratory animals and, furthermore, that particular individuals could be as much as 10-fold more sensitive than the average person. Since then, numerous assessments of the risk posed by dioxins and dioxin-like PCBs to human health have been performed and various models, including breastfeeding infants, employed to estimate human exposure to POPs.[5,6] Finally, the WHO set exposure limits within the range of 1–4 pg WHO-TEQ kg^{-1} bw d^{-1} (WHO-TEQ = World Health Organization toxic equivalent; bw = body weight),[7] with the upper value being considered to be "maximal tolerable intake on a provisional basis" and the ultimate goal to reduce human intake below 1 pg WHO-TEQ kg^{-1} bw d^{-1}.

Food is generally recognized as the source of 90% of total human daily intake of PCDD/Fs and dioxin-like PCBs.[8] For instance, the dioxin content of fatty fish from the Baltic Sea often exceeds the prescribed maximum limits,[9,10] so that the levels of these compounds detected in individuals who frequently consume such fish are comparable with those inhabitants of Seveso, Italy, following the accidental release of 2,3,7,8-tetrachlorodibenzo-*p*-dioxin (2,3,7,8-TCDD).[11]

Therefore, in attempt to reduce the intake of these substances, the Swedish National Food Administration has recommended that herring, salmon and sea trout from the Baltic Sea and Arctic char from Lake Vättern should be consumed no more than once a week, on average.[4] Moreover, bans have been imposed on eating cod liver from the Baltic Sea and from coastal waters around Gothenburg, which contain very large quantities of DDT and PCBs, respectively. Since fish liver has never been consumed in large amounts in Baltic countries, the main routes of exposure to PCBs and dioxin-like compounds here are dairy and meat products. Obviously, during infancy, breast and formula milk are the major food sources of these chemicals.

In various organisms, including human beings, considerable concentrations of these pollutants have been detected in blood and adipose tissue and they also cross the placenta and accumulate in breast milk. Since exposure both before and after birth has already given rise to subtle abnormalities in approximately 10% of the newborn infants in the Netherlands,[12] pregnant Dutch mothers are advised to reduce their consumption of cow's milk and milk products, as well as fish. A recent study indicates that a diet high in fat enhances dioxin-like activity and exposure *in utero* may disturb the early development of the embryo and fetus. Furthermore, maternal consumption of fish from areas contaminated with POPs enhances the risk of prenatal exposure to the most heavily chlorinated PCB homologs.[13] It has also been proposed that levels of persistent bioaccumulating toxicants in the placenta and breast milk can be used as indicators of body burden.[14] The observation that serum levels of PCBs and DDE in children correlate positively with the length of breast-feeding points to breast milk as the major source of these pollutants in young individuals.

Obviously, infants are particularly vulnerable to at least certain food chemicals and cannot be considered to be "small adults". Their nervous, respiratory and reproductive systems are not yet fully developed and they are less able to excrete certain toxins. At the same time, they sometimes generate lower levels of toxic metabolites.[15] Moreover, infants tend to be exposed to relatively high levels of food chemicals, since they consume more food per kilogram body weight.

Although the most pronounced dietary exposure to dioxins is encountered by breast-fed infants, there is considerable evidence that the concentrations of dioxins in breast milk have fallen in recent years.[16] In addition, since breast-feeding has a measurable positive influence on immunological development, a formula diet should not be recommended as a means of lowing dioxin intake. Furthermore, neither prenatal exposure to PCBs nor postnatal exposure to PCBs and dioxins appears to alter the neurological status at 42 months of age.[17]

At the same time, according to statistics from 2007, few European women breastfeed their infants exclusively at six months of age and only 33% of infants in the USA are exclusively breastfed up to 3 months of age.[18,19] The increasing number of mothers who feed their babies industrially processed formula milk, or solids such as vegetable and meat or fish purée, has caused the baby food market to grow significantly and, with it, the assortment of products offered. Most investigations to date have focused on POPs in breast milk and only a few have examined PCDD/Fs, PCBs and organochlorine pesticides (OCPs) in formula milk consumed by infants.[20–24] Therefore, the CASCADE research consortium (Chemicals as contaminants in the food chain; EU Network of Excellence, under FP6: Targeting health risks in food) decided to evaluate the present dietary exposure levels of non-breast-fed European infants.

2.2 Analysis of Baby Foods

2.2.1 Food Items Investigated

A market basket of commercial baby foods designed for consumption during the first 9 months of life by an "average" EU baby fed with infant formulae, and weaned (at the 5th month) with industrial solid foods and beverages, has been put together.[25] The subsequent analysis concerned an infant whose diet consisted entirely of commercial baby foods, *i.e.* who was not breast-fed at all. Therefore, the diet basket was modified monthly and involved six different types of infant formulae, *i.e.* the "starting" and "follow-on" varieties of milk, soy and hypoallergenic (HA) formulae. Solid foods and beverages were introduced progressively, starting from month 5 to month 9 month of age, with modification by introducing a new food every month.

The 2007 market data made available by "Food for Thought" (www.fft.com) was used to identify the major brands of baby food, and their shares of the market in 22 EU countries were considered to represent the entire EU. On the basis of this modification, products of 42 different infant formulae were obtained from six different countries (France, Germany, Italy, Portugal, Sweden and the UK), along with products of a random selection of 22 different solid food and beverage products from five different countries (Germany, Italy, France, Spain and the UK).

2.2.2 Quantification of POPs and Estimation of Daily Consumptions

Six pooled samples each of "starting" infant formulae of milk-based (Mf), soy-based (Sf) and hypoallergenic-based (HAf) and "follow-on" infant formulae of milk-based (fMf), soy-based (fSf) and hypoallergenic-based (fHAf) and five pooled samples of the solid food items and beverages (SFB) were prepared in such a manner that the proportion of each product in the pooled sample was as its share of the European market. Detail information concerning sample preparation, the clean-up procedure and instrumental parameters are provided elsewhere.[26] The tetra- to octa-PCDDs and PCDFs and tetra- to hepta-PCBs were identified and quantified in pg WHO-TEQ g^{-1} fresh weight (fw) (TEQ 1998 values used) and pentachloro[$^{13}C_6$]benzene , α-[$^{13}C_6$]HCH , γ-[$^{13}C_6$]HCH, β-[$^{13}C_6$]HCH, δ-[$^{13}C_6$]HCH, ε-[$^{13}C_6$]HCH, hexachloro[$^{13}C_6$]benzene, [$^{13}C_6$]-heptachlor, [$^{13}C_{12}$]aldrin, oxy[$^{13}C_{10}$]chlordane, [$^{13}C_{10}$]heptachlorepoxide, 2,4′-[$^{13}C_{12}$]DDE, 4,4′-[$^{13}C_{12}$]DDE, *trans*-[$^{13}C_{12}$]chlordane, 4,4′-[D_8]DDD, [$^{13}C_{12}$]dieldrin,

2,4′-[^{13}C$_{12}$]DDT, 4,4′-[^{13}C$_{12}$]DDT, [^{13}C$_{12}$]methoxychlor and [^{13}C$_{10}$]mirex were identified and quantified in pg g^{-1} fw. The analytical laboratory involved is quality assured according to DIN EN ISO/IEC 17025 and accredited for the analysis of PCBs, PCDD/Fs and OCPs.

The mean daily dietary exposure to PCDD/Fs, PCBs and OCPs was then calculated on the basis of these levels and estimates of the average amount of "infant formulae" (mL d^{-1}) and "solid food and beverages" (g d^{-1}) consumed by an "average" baby in the EU. The values thus obtained in g kg^{-1} body weight (bw) were 25.2, 22.9, 21.0 and 18.3 in the "starting" formulae consumed by infants 0–1, 1–2, 2–3 and 3–4 months of age, respectively; 12.8, 8.0, 4.6, 4.4 and 4.2 for the "follow-on" formulae consumed by infants 4–5, 5–6, 6–7, 7–8 and 8–9 months of age, respectively; and 28.8, 54.7, 72.6, 70.2 and 71.9 for the solid foods and beverages consumed by infants 4–5, 5–6, 6–7, 7–8 and 8–9 months of age, respectively.

2.2.3   Levels of POPs in Commercial Baby Foods

Table 2.1 presents the lower bound (LB) and upper bound (UB) concentration levels of PCBs and PCDD/Fs in infant formulae and solid foods and beverages samples. No or negligible levels of PCDD/Fs or PCBs were detected in the solid foods and beverages, while the infant formulae exhibited an LB e of 0.01–0.05 (WHO-TEQ) pg g^{-1} fw, corresponding to 0.04–0.2 (WHO-TEQ) pg g^{-1} lipid (the package labeled lipid content of approximately 25 g per 100 g fw).

Among the different isomers of HCH, α-HCH was detected only in Sf and fHAf and β-HCH only in fHAf (Table 2.2). The α-, β- and γ-isomers of HCH were present in all solid foods and beverages samples, with a monthly stepwise increase in the content of the α-HCH as the diet became more and more varied. The most biologically active isomer of HCH, the γ-isomer (lindane), was found predominantly in the solid foods and beverages samples, although still at levels much lower than the maximum residue limits (MRL) proposed for this compound (*i.e.* 0.01 mg kg^{-1} fat).[27]

As also given in Table 2.2, in almost all of the samples of baby food analyzed, DDTs were present at levels above the detection limit. In addition, in both types of baby foods (*i.e.* infant formulae and solid foods and beverages) the levels of 4,4′-DDT, 2,4′-DDT, 4,4′-DDD, 2,4′-DDD and 2,4′-DDE were relatively similar. Out of the six DDTs analyzed 4,4′-DDE was present at highest levels 20.9 pg g^{-1} fw in the milk formulae and SFB, where the corresponding levels of SFB8 and SFB9 were 45.5 and 37 pg g^{-1} fw, respectively. On a lipid basis these latter values were 0.5 and 0.5 ng g^{-1}, respectively, which is far lower than the levels of 4,4′-DDE values in breast milk reported from Australia (279 ng g^{-1} lipid),[28] Taiwan (228 g g^{-1} lipid),[29] the Czech Republic (600 ng g^{-1} lipid)[30] and China (1087 ng g^{-1} lipid).[31]

Heptachlor, *cis*-chlordan, *trans*-heptachlorepoxide and endrin were not detected in any of the samples assayed (Table 2.2). However, *trans*-chlordane, oxychlordane and aldrin were present at negligible concentrations in solid

Table 2.1 Lower bound (LB) and upper bound (UB) levels of PCDD/Fs and PCBs (WHO-TEQ pg g^{-1} fw) in samples of infant formula milk and solid foods and beverages.[a]

Type of baby food		Mf	Sf	HAf	fMf	fSf	fHAf	Solid baby foods and beverages				
Infant age (months)		0–4th month			Up to the 5th month			5th	6th	7th	8th	9th
PCDD/Fs (WHO-TEQ pg g^{-1} fw)	Lower bound (LB)	0.04	0.05	0.11	0.04	0.02	0.01	n.d.	0.0035	0.00001	0.00002	0.0005
	Upper bound (UB)	0.09	0.09	0.11	0.09	0.07	0.12	0.03	0.0436	0.0384	0.0640	0.0693
PCBs (WHO-TEQ pg g^{-1} fw)	Lower bound (LB)	0.0010	0.0003	0.0005	0.0030	0.0001	0.0006	0.0000	0.01	0.01	0.01	0.01
	Upper bound (UB)	0.0021	0.0013	0.0006	0.0031	0.0013	0.0017	0.17	0.01	0.01	0.01	0.01

[a] Mf = "starting" infant formula, milk based; Sf = "starting" infant formula, soy based; HAf = "starting" infant formula, hypoallergenic based; fMf = "follow-on" infant formula, milk based; fSf = "follow-on" infant formula, soy based; fHAf = "follow-on" infant formula, hypoallergenic-based. n.d. = no data.

Table 2.2 OCP levels (pg g^{-1} fw) in samples of infant formula milk and solid foods and beverages.[a]

Substance	Mf	Sf	HAf	fMf	fSf	fHAf	SF5	SF6	SF7	SF8	SF9
α-HCH	n.d.	2.7	n.d.	n.d.	n.d.	2.3	4.3	6.6	6.9	8	7.6
β-HCH	n.d.	n.d.	n.d.	n.d.	n.d.	1.7	1.1	0.9	1.5	1.4	1.2
γ-HCH	n.d.	n.d.	n.d.	n.d.	n.d.	n.d.	22.7	10.7	13.3	17.4	13.4
δ-HCH	n.d.	n.d.	n.d.	n.d.	n.d.	n.d.	n.d.	n.d.	n.d.	n.d.	n.d.
ε-HCH	n.d.	n.d.	n.d.	n.d.	n.d.	n.d.	n.d.	n.d.	n.d.	n.d.	n.d.
Pentachlorobenzene	n.d.	4.2	7.3	n.d.	n.d.	6.4	0.28	3.7	8.2	27.4	122
Hexachlorobenzene	n.d.	69.9	60.8	n.d.	n.d.	45.9	n.d.	n.d.	n.d.	0.52	8.9
4,4′-DDT	n.d.	5.8	1.3	7.6	8.6	1.8	2.1	1.9	2.2	5.6	3
2,4′-DDT	2.9	1.6	n.d.	2.7	1	0.84	0.9	0.8	0.9	1.9	1.1
4,4′-DDD	4.2	3.2	5.2	10.8	5.3	3.1	0.7	0.7	1.0	4.6	3.9
2,4′-DDD	2.4	0.84	1.5	2.4	1.7	0.84	0.1	0.1	0.1	1.3	0.9
4,4′-DDE	20.9	1.9	10.1	16.5	4.9	11.2	8.6	10.1	9.4	45.5	37
2,4′-DDE	4	0.67	0.61	n.d.	0.72	0.84	n.d.	n.d.	n.d.	0.3	0.2
trans-Chlordane	n.d.	1.8	n.d.	n.d.	n.d.	n.d.	0.7	0.3	0.3	0.3	0.4
cis-Chlordane	n.d.	n.d.	n.d.	n.d.	n.d.	n.d.	n.d.	n.d.	n.d.	n.d.	n.d.
oxy-Chlordane	n.d.	n.d.	n.d.	n.d.	n.d.	n.d.	n.d.	n.d.	0.4	0.4	0.3
Heptachlor	n.d.	n.d.	n.d.	n.d.	n.d.	n.d.	n.d.	n.d.	n.d.	n.d.	n.d.
cis-Heptachlorepoxide	10.9	9.7	2.2	23	11.2	1.5	1.3	1.6	1.5	1.4	1.3
trans-Heptachlorepoxide	n.d.	n.d.	n.d.	n.d.	n.d.	n.d.	n.d.	n.d.	n.d.	n.d.	n.d.
Aldrin	n.d.	n.d.	n.d.	n.d.	n.d.	n.d.	n.d.	n.d.	n.d.	0.2	0.2
Dieldrin	8.5	27	3.9	18	25.2	6.4	5.7	8.1	7.0	4.7	4.2
Endrin	n.d.	n.d.	n.d.	n.d.	n.d.	n.d.	n.d.	n.d.	n.d.	n.d.	n.d.
Endosulfan-I	74.5	34.3	101	89.9	33.9	116	15.3	22.4	17.8	19.5	16
Endosulfan-II	139	39.6	132	83.9	61.8	148	35.5	37	27.4	33.2	24.9
Methoxychlor	n.d.	0.82	n.d.	n.d.	1.4	n.d.	0.4	0.3	0.5	0.6	0.5
Mirex	n.d.	n.d.	1.4	n.d.	n.d.	n.d.	n.d.	n.d.	n.d.	n.d.	n.d.

[a]Mf, "starting" infant formula, milk based; Sf, "starting" infant formula, soy based; HAf, "starting" infant formula, hypoallergenic based; fMf, "follow-on" infant formula, milk based; fSf, "follow-on" infant formula, soy based; fHAf, "follow-on" infant formula, hypoallergenic based; SFB5, SFB6, SFB7, SFB8, SFB9, solid foods and beverages designed for 5-, 6-, 7-, 8- and 9-month-old infants, respectively. n.d. = no data.

foods and beverages, in particular food designed for 8- and 9-month-old infants. Although *cis*-heptachlorepoxide was found in all the baby foods, with the highest level of 23 pg g^{-1} fw (0.09 ng kg^{-1} lipid) in "follow-on" milk formula, its levels were well below the MRL of 0.006 mg kg^{-1} lipid.[27]

2.2.4 Daily Exposure to POPs of 0–9 Months of Age Non-breast-fed Infants

When exposure of 0- to 9-month-old infants *via* commercial baby food was calculated (Table 2.3), this exposure appeared to decrease from month to month, owing to the reduced consumption of formula per kg body weight. Moreover, exposure to dioxin-like PCBs *via* the six types of infant formula during the first nine months of life was far lower than the exposure to PCDD/ Fs (Table 2.3). Exposure of 5- to 9-month-old infants to PCBs *via* the solid foods and beverages investigated was estimated to be between 0 (n.d.) and 0.73 pg WHO-TEQ kg^{-1} bw d^{-1}, which is higher than the corresponding value for PCDD/Fs from the fifth to ninth months of postnatal life.

The highest estimated exposure to PCDD/Fs (more than 2 pg WHO-TEQ kg^{-1} bw d^{-1}) was exhibited by infants consuming "starting" hypoallergenic formula who had an average body weight of less than 6.4 kg. In these cases, the average monthly dietary exposure may nearly reach the level or even exceed the

Table 2.3 Estimated dietary exposure to PCDD/Fs and PCBs (pg WHO-TEQ kg^{-1} bw d^{-1}) of 0–9 months of age non-breast-fed infants fed with baby foods available on the EU market.

	Infant age (months)								
Type of baby food	*0–1*	*1–2*	*2–3*	*3–4*	*4–5*	*5–6*	*6–7*	*7–8*	*8–9*
PCDD/Fs TEQ (pg kg^{-1} bw d^{-1})									
Milk infant formula	1.00	0.92	0.84	0.73	0.51	0.32	0.18	0.18	0.17
Soy infant formula	1.26	1.15	1.05	0.91	0.26	0.16	0.09	0.09	0.09
Hypoallergenic infant formula	2.78	2.52	2.31	2.01	0.13	0.08	0.05	0.04	0.04
Solid foods and beverages					0.00	0.19	0.00	0.00	0.04
PCBs TEQ (pg kg^{-1} bw d^{-1})									
Milk infant formula	0.03	0.02	0.02	0.02	0.04	0.02	0.01	0.01	0.01
Soy infant formula	0.01	0.01	0.01	0.01	0.00	0.00	0.00	0.00	0.00
Hypoallergenic infant formula	0.01	0.01	0.01	0.01	0.01	0.00	0.00	0.00	0.00
Solid foods and beverages					0.00	0.55	0.73	0.70	0.72
Total TEQ (pg kg^{-1} bw d^{-1})									
Milk infant formula	1.03	0.94	0.86	0.75	0.55	1.08	0.92	0.89	0.94
Soy infant formula	1.27	1.16	1.06	0.92	0.26	0.90	0.82	0.79	0.84
Hypoallergenic infant formula	2.79	2.53	2.32	2.02	0.14	0.82	0.77	0.75	0.80

recommended 70 pg WHO-TEQ kg bw month^{-1}. For example, for 0–1- and 1–2-month-old infants fed "starting" hypoallergenic formula, the estimated monthly intake is 84 and 76 pg WHO-TEQ kg^{-1} bw month^{-1}, respectively. Furthermore, consideration of UB values may suggest critical dietary exposure for both groups of infants consuming milk-based and soy-based "starting" infant formulae. Although there were similar UB levels for PCDD/Fs in "starting" and "follow-on" formulae of this kind, lower than 2 pg WHO-TEQ kg^{-1} bw d^{-1} exposure was estimated during the weaning period due to differences in the pattern of consumption.

In general, excluding hypoallergenic formulae, the baby foods may expose 0–4- and 4–9-month-old non-breast-fed infants to approximately 1 and 0.7 pg WHO-TEQ kg^{-1} bw d^{-1}, respectively, values that when expressed by month do not exceed the recommended PMTI (Provisional Monthly Tolerable Intake). Similarly, the estimated chronic exposure to PCDD/Fs and dioxin-like PCBs was 1.1 pg WHO-TEQ kg^{-1} bw d^{-1} for infants 5 months of age who consume formulae and vegetables available on the Dutch market.[3] Overall, the daily intakes estimated here do not exceed the maximum recommended TDI of 4 pg WHO-TEQ kg^{-1} bw d^{-1} and, moreover, are orders of magnitude lower than the estimated total TEQ daily intake of 117 and 271 pg kg^{-1} bw d^{-1} for breast-fed infants.[32]

In general, only low levels of OCPs were detected in the samples of baby food investigated. The highest daily intake of 0.0001 μg DDT kg^{-1} bw d^{-1} was estimated for 4- to 5-month-old infants fed fMf, a value below the recommended limit of 10 μg kg bw d^{-1}.[27] With regard to α-, β- and γ-HCH, the worst-case scenario would involve exposure of 0.6, 0.02 and 0.6 ng kg^{-1} bw d^{-1} for infants consuming solid foods and beverages during the eighth month of life (SFB8), fHAf during months 4 and 5 and solid foods and beverages during the fifth month (SFB5), respectively. However, all of these values are also negligible compared to the acceptable daily intakes (ADI) of 5, 1 and 12.5 μg kg^{-1} bw d^{-1}, respectively.[33,34]

2.2.5 Comparison of Dietary Exposure of Infants to POPs *via* Formula and Breast Milk

In 51 samples of human breast milk recently collected from five cities throughout Turkey the average level of PCDD/Fs was 7.5 WHO-TEQ pg g^{-1} lipid (with 100 mL of milk containing on average 3.1 g of lipid).[25] On the basis of the daily consumption in the form of liquid infant formula calculated here and average body weight, the estimated dietary exposure of Turkish infants to PCDD/Fs would be 39.4, 35.4, 33, 28.6, 18.9, 11.9, 6.9, 6.6 and 6.4 pg WHO-TEQ kg^{-1} bw d^{-1} during 0–1, 1–2, 2–3, 3–4, 4–5, 5–6, 6–7, 7–8 and 8–9 months of postnatal life, respectively. In Korea, an infant has been predicted to ingest as much as 85 pg WHO-TEQ kg^{-1} bw d^{-1} during the first year of life,[35] and Päpke reported a corresponding intake of 24–145 pg WHO-TEQ kg^{-1} bw d^{-1} for 5 kg babies.[36] In general, most investigations have demonstrated that the average dietary exposure of breast-fed infants to PCDD/Fs is considerably higher than the TDI recommended by the WHO and, moreover, these levels are much higher than those estimated here. In interpreting such findings it should be remembered that the TDI is designed for comparisons of dietary exposure over an entire lifetime and a short period of higher dietary exposure may not lead to health problems.

2.3 Conclusions and Future Outlook

The present assessment of the levels of PCDD/Fs, PCBs and OCPs, three classes of POPs, in infant formulae and solid foods and beverages designed for infants focused on the products with the largest shares of the market in 22 EU countries. Overall, the PCDD/Fs and PCBs in baby food were low, with the highest value of 0.11 pg g^{-1} (0.44 pg g^{-1} lipid) being detected in "starting" hypoallergenic formulae. In general, the levels of PCDD/Fs in cows' milk is higher than the level in formula milk determined here,[15] presumably because reconstruction of infant formula from cows' milk involves intentional removal of most lipids of animal origin, in which dioxin accumulates. However, owing to the relatively high food consumption of infants per kilogram body weight, the low levels of PCDD/F and PCB concentrations observed here in baby foods may result in an intake of more than 1 pg WHO-TEQ kg^{-1} bw d^{-1} by non-breast-fed infants during their first 9 months of life. Indeed, infants 0–4 months of age who consume "starting" hypoallergenic formula may have a daily intake of greater than 2 pg WHO-TEQ kg^{-1} bw d^{-1}.

Among the large number of OCPs analyzed here, only some could be detected and even their levels were negligible in comparison to the established MRL for food. Thus, the estimated daily exposure of non-breast-fed infants to OCPs through baby foods investigated appears not to be harmful to health.

Finally, the levels of contaminants detected in major baby food products sold in the EU and the estimated daily exposure for non-breast-fed infants are magnitudes lower than those reported for nursing infants. The intake of these chemicals by women must be lowered years before they become pregnant.[12] Up to the age of 10, Swedish children have a median TEQ intake greater than the TDI of 2 pg TEQ kg^{-1} body weight, owing to exposure *via* dairy and fish products.[15] The levels of POPs, including dioxins and PCBs, in food should provide a basis for recommendations concerning the risks posed by commercial infant food to the health of young children.

Acknowledgements

The authors would like to thank all CASCADE partners who assisted in the acquisition of information and shopping of baby food products, namely Ingemar Pongratz, Lars-Arne Haldosen and Jean-Pierre Cravedi.

The study was financial supported by the European Union network CASCADE (FOOD-CT-2003-506319) within the frame of WP19 projects (bread project and baby food project).

References

1. USEPA, report no. EPA/600/P-01/001, US Environmental Protection Agency, Washington, 2002.
2. Deutschen Forschungsgemeinschaft, *Hormonally Active Agents in Food: Symposium*, ed. G. Eisenbrand, Wiley-VCH, Weinheim, 1998.
3. P. J. M. Weijs, M. I. Bakker, K. R. Korver, K. Goor Ghanaviztchi and J. H. Wijnen, *Chemosphere*, 2006, **64**, 1521.
4. C. Bernes and M. Naylor, *Persistent Organic Pollutants*, Swedish EPA, Stockholm, 1999.
5. S. Trapp, L. M. Bomholtz and C. N. Legind, *Environ. Pollut.*, 2008, **156**, 90.
6. J. A. Arnot, D. Mackay, T. F. Parkerton, R. T. Zaleski and C. S. Warren, *Environ. Toxicol. Chem.*, 2010, **29**, 45.
7. WHO, *Food Add. Contam.*, 2000, **17**, 223.
8. T. Tsutsumi, T. Yanagi, M. Nakamura, Y. Kono, H. Uchibe, T. Iida, T. Hori, R. Nakagawa, K. Tobiishi, R. Matsuda, K. Sasaki and M. Toyoda, *Chemosphere*, 2001, **45**, 1129.
9. EU, EC no. 2375/2001, Council of the European Communities, Brussels, 2001.
10. EU, 2002/201/EC, Commission of the European Communities, Brussels, 2002.
11. H. Kiviranta, T. Vartiainen, M. Verta, J. T. Tuomisto and J. Toumisto, *Lancet*, 2000, **355**, 1883.
12. J. G. Koppe, *Eur. J. Obstet. Gynecol. Reprod. Biol.*, 1995, **61**, 73.
13. P. Stewart, T. Darvill, E. Lonky, J. Reihman, J. Pagano and B. Bush, *Environ. Res. Sect. A*, 1999, **80**, 87.
14. H. Shen, K. M. Main, H. E. Virtanen, I. N. Damggard, A.-M. Haavisto, M. Kaleva, K. A. Boisen, I. M. Schmidt, M. Chellakooty, N. E. Skakke-baek, J. Toppari and K.-W. Schramm, *Chemosphere*, 2007, **67**, 256.
15. C. Bergkvist, M. Öberg, M. Appelgren, W. Becker, M. Aune, E. H. Ankarberg, M. Berglung and H. Håkansson, *Food. Chem. Toxicol.*, 2008, **46**, 3360.
16. S. Lorán, S. Bayarri, P. Conchello and A. Herrera, *Chemosphere*, 2007, **67**, 513.
17. C. I. Lanting, S. Patandin, V. Fidler, N. Weisglas-Kuperus, P. J. J. Sauer, E. R. Boersma and B. C. L. Touwen, *Early Hum. Dev.*, 1998, **50**, 283.
18. EU project on promotion of breastfeeding in Europe: http://europa.eu.int/comm/health/ph_projects/2002/promotion/promotion_2002_18_en.htm.
19. CDC national immunization survey data; breastfeeding among U.S. children born 1999–2006: http://www.cdc.gov/breastfeeding/data/NIS_data/index.htm.
20. J.-F. Hsu, Y. L. Guo, C.-H. Liu, S.-C. Hu, J.-N. Wang and P.-C. Liao, *Chemosphere*, 2007, **66**, 311.
21. J. Chovancová, A. Kočan and S. Jursa, *Chemosphere*, 2005, **61**, 1305.
22. L. Ramos, M. Torre, F. Laborda and M. L. Marina, *J. Chromatogr. A*, 1998, **823**, 365.

23. T. L. Cederberg, S. Sørensen, K. H. Lund and S. Friis-Wandall, in *Organohalogen Compounds, Proceedings of the 30th International Symposium on Halogenated Persistent Organic Pollutants*, San Antonio, TX, 2010, pp. 1430–1433.
24. A. Schecter, P. Fürst, C. Fürst, H.-A. Meemken, W. Groebel and D. C. Vu, *Chemosphere*, 1989, **19**, 913.
25. R. Piccinelli, M. Ferrari, K.-W. Schramm, M. Pandelova and C. Leclercq, *Food Add. Contam. A*, 2010, **27**, 1337.
26. I. Cok, M. K. Donmez, M. Uner, E. Demirkaya, B. Henkelmann, H. Shen, J. Kotalik and K.-W. Schramm, *Chemosphere*, 2009, **76**, 1563.
27. FAO/WHO, *Codex Maximum Limits for Pesticide Residues*, Codex Alimentarius Commission, FAO and WHO, Rome, 2006.
28. J. F. Mueller, F. Harden, L.-M. Toms, R. Symons and P. Fürst, *Chemosphere*, 2008, **70**, 712.
29. H.-R. Chao, S.-L. Wang, T.-C. Lin and X.-H. Chung, *Chemosphere*, 2006, **62**, 1774.
30. M. Cerna, V. Bencko, M. Brabec, J. Smid, A. Krskova and L. Jech, *Chemosphere*, 2010, **78**, 160.
31. G. Zhao, Y. Xu, W. Li, G. Han and B. Ling, *Sci. Total Environ.*, 2007, **378**, 281.
32. V. Bencko, M. Cerna, L. Jech and J. Smid, *Environ. Toxicol. Pharmacol.*, 2004, **18**, 83.
33. WHO/FAO, Joint Meeting on Pesticide Residues, *Pesticide Residues in Food 2002, Part II, Toxicological Evaluations*, WHO/PCS/03.1, World Health Organization, Geneva, 2003, pp. 117–164.
34. Deutschen Forschungsgemeinschaft, *Hexachlorcyclohexan-Kontamination-Ursachen, Situation und Bewertung*, Kommission zur Prüfung von Rückständen in Lebensmitteln, Mitteilung IX, Boldt Verlag, Boppard, 1982.
35. J. Yang, D. Shin, S. Park, Y. Chang, D. Kim and M. G. Ikonomou, *Chemosphere*, 2002, **46**, 419.
36. O. Päpke, *Organohalogen Compds.*, 1999, **44**, 5–8.

Chemicals Targeting the Reproductive Axis

K. SVECHNIKOV AND O. SÖDER

Department of Women's and Children's Health, Pediatric Endocrinology Unit, Karolinska Institute & University Hospital, Q2:08, SE-17176 Stockholm, Sweden

3.1   Introduction

Androgens play a critical role in the development of the normal male phenotype. Disruption of androgen biosynthesis and actions by environmental endocrine disrupting compounds (EDCs), of anthropogenic or natural origin, may influence critical cellular processes controlling steroidogenesis by Leydig cells (*e.g.* transport and delivery of cholesterol into mitochondria, steroidogenic enzyme expression or activity) and androgen binding to the androgen receptor (AR). Such actions may result in incomplete masculinization and malformations of the male reproductive tract of mammals, including humans. Androgen deficiency during critical developmental time windows of prenatal development is thought to disturb differentiation of male external genitalia and the Wolffian duct,[1] a process crucial to the proper development of male internal reproductive organs. An increased incidence of human male reproductive developmental disorders, such as cryptorchidism (failure of the testis to descend into the scrotum during fetal development), hypospadia (a birth defect of the urethra development, resulting in aberrant location of the external orifice) and testicular cancer, as well as decreased quality of semen, has been proposed to be due to the action of EDCs.[1] It has been hypothesized that EDCs play a role in the etiology of these abnormalities, collectively termed testicular dysgenesis

Issues in Toxicology No. 11
Hormone-Disruptive Chemical Contaminants in Food
Edited by Ingemar Pongratz and Linda Vikström Bergander
© Royal Society of Chemistry 2012
Published by the Royal Society of Chemistry, www.rsc.org

syndrome (TDS).[1] Reliable data on the prevalence of TDS are scarce but >5% of the Danish male population was reported to suffer from undescended testes, hypospadia or testicular cancer.[2] In addition, the incidences of cryptorchidism and hypospadias at birth were reported to reach 2–9% in Nordic countries.[2] These disturbances in the development of male reproductive organs may be associated with dysfunction of fetal Leydig cells and deficiency in androgen production or action. This suggestion is supported by several recent studies in the rat, showing that inhibition of androgen production within the masculinization programming window increases the risk of cryptorchidism, hypospadias and reduced penis size at puberty or in adulthood.[3,4] Hypothetically, the increased incidence of malformations of the male reproductive tract may be associated with enhanced day-to-day exposure of humans to a number of widespread environmental chemicals that have the capacity to inhibit testosterone production by fetal Leydig cells.

3.2 The Role of Androgens in Male Fetal Sex Differentiation

Development of the male reproductive tract is a dynamic process requiring the interaction of numerous factors and hormones. Among the major factors essential to the proper development of the internal and external male reproductive organs are the androgens testosterone and dihydrotestosterone (DHT).[5] Masculinization of the reproductive tract involves the differentiation of the internal (epididymis, vas deferens, seminal vesicles and prostate) and external (penis, scrotum and perineum) genitalia.[6] In humans, testosterone reaches its maximal values between 11 and 18 weeks of gestation and stimulates differentiation of the Wolffian duct into the epididymis, vas deferens and seminal vesicles. Masculinization of the external genitalia and prostate is mediated primarily by DHT, a more potent metabolite of testosterone produced by the enzyme 5α-reductase (5αR).[5] Simultaneously, the Sertoli cells of the fetal testis secrete anti-Müllerian hormone (AMH), which induces regression of the Müllerian duct.[5] The actions of both testosterone and DHT are mediated by AR that is expressed in both the internal and the external genitalia. In the human male fetus, AR is expressed as early as after 8 weeks of gestation, prior to the onset of testicular androgen secretion, and at a higher level in genital than in urogenital structures. Upon high-affinity binding of its agonists, AR undergoes conformational changes and is imported into the nucleus, where it dimerizes.[7] This ligand dimer complex then binds to an androgen response element located within introns or flanking regions of responsive genes and activates their transcription.[8] The resulting gene products are responsible for androgen-dependent functions of male sexual differentiation.

The main androgen-dependent stages of differentiation of the reproductive tract structures occur during the masculinization programming window, while some organs (*e.g.* the penis) continue the growth under influence of androgens postnatally.[9] It is well known that insufficient androgen production or action during the masculinization programming window causes disorders of masculinization such as hypospadias, cryptorchidism and defects in the prostate development, as well as reduced anogenital distance (AGD) and penis length in rodents.[3] However, suppression of androgen action after the programming window, when both luteinizing hormone (LH) and its receptor are already expressed, does not affect the masculinization process.[3]

3.3 Fetal Leydig Cells

During the prenatal period, the fetal Leydig cell (FLC) is the primary source of androgens. These cells secrete testosterone and other androgens, which regulate not only the masculinization of internal and external male genitalia, but also neuroendocrine functions. The development of fetal Leydig cells can be separated into three stages, *i.e.* differentiation, fetal maturation and involution.[10] FLCs start to appear in the mesenchyme of the developing prenatal testis at embryonic day (E) 12.5 in mice, 14.5 in rats and 7–8 weeks of gestation in human. In the fetal human testis, differentiation occurs at a gestational age of 7–14 weeks, maturation during the weeks 14–18 of gestation and the involution thereafter until the time of full-term birth.

In the male fetus, Leydig cells seem to arise from multiple sources, including the coelomic epithelium, gonadal ridge mesenchyme and migrating mesonephric cells.[11,12]

In humans, the LH receptor (LHR)-dependent testosterone production by FLC becomes substantial at fetal age of week 10, when the LHR start to express.[13] However, testosterone production by human FLC at very early fetal ages (7 –10 weeks) is suggested to be partially or completely LH/hCG independent.[6] Further experiments with human fetal testis explants have demonstrated LH/hCG responsiveness of steroidogenesis at 7–12 weeks of gestation, while human fetal testis explants at 7 weeks of gestation produce testosterone in response to retinoic acid but not LH/hCG treatment.[14] In contrast to humans, the rat does not produce chorionic gonadotropin (HG).[15] Despite the earlier presence of LHR (E14.5–E15.5), LH secretion in rats does not start until E17.5.[16] Moreover, in a mice model with targeted inactivation of the LHβ subunit gene, the males exhibit normal prenatal masculinization.[17] These observations provide strong support for the idea that steroidogenesis in rat FLC, in contrast to human ones, are more dependent on paracrine factors before the onset of LH secretion.[6]

FLCs possess a well-developed steroidogenic machinery expressing the key steroidogenic enzymes (*e.g.* StAR, P450scc, 3β-HSD, P450c17, 17β-HSD and 5αR) required for androgen synthesis. Testosterone secreted by fetal Leydig cells is required for masculinization and proper development of male reproductive organs. This steroid produced by FLCs is converted to DHT by 5αR.[18] DHT is a more potent androgen than testosterone and controls proper development of male external genitalia, including the penis and scrotum. Moreover, fetal Leydig cells also produce insulin-like factor 3 (Insl3), a protein that plays a critical role in the regulation of the early phase of testicular descent.[19] Thus, FLC dysfunction induced by endogenous or exogenous factors may lead to incomplete masculinization and various malformations in the male reproductive tract of humans and animals (*e.g.* cryptorchidism, hypospadia, micropenis). Therefore, one can assume that environmental chemicals of natural or anthropogenic origin that can affect androgen production by FLCs may significantly influence the masculinization of the male fetus and proper development of the reproductive tract.

3.4   Characteristics of Endocrine Disrupting Chemicals

Many compounds introduced into the environment by human activity are capable of disrupting the endocrine system of animals and humans. The hormonally regulated reproductive system can be permanently affected by a wide range of exogenous compounds able to interfere with the hormonal signaling at various levels. Such interaction may affect the development and function of the reproductive organs in many animal species. The endocrine disrupting chemicals (EDCs) are a heterogeneous group of xenobiotics which can mimic natural hormones, inhibit the action of hormones or alter their biosynthesis and/or catabolism. EDCs include groups of pesticides (*e.g.* vinclozolin, linuron, procymidone), persistent organic pollutants (DDT and metabolites, dioxins), industrial chemicals (*e.g.* alkyl phenols, phthalates, bisphenol A) and some metals (*e.g.* arsenic or cadmium), as well as some natural compounds such as phytoestrogens. The reproductive system represents one of the main targets of EDCs as *in utero* exposure to environmental pollutants has been associated with increased incidences of reproductive organ malformations (*e.g.* hypospadias, cryptorchidism) and testicular cancer in humans,[6,20,21] as well as a decline in sperm counts in some areas.[22,23] The risk of exposure of human fetuses to EDCs from food and different man-made chemicals is currently significantly increased. For example, high concentrations of phthalate metabolites were found in the urine of women of childbearing age,[24] a finding that can be associated with extensive use of personal care products (*e.g.* skin creams and nail polish) containing phthalates. Moreover, measurement of environmental contaminants in amniotic fluid samples have revealed that about 30% of fetuses in the Los Angeles area are exposed to EDCs *in utero*, the consequences of which remain unknown at this time.[25] The timing and duration of exposure as well as the EDC dose determine the cascade of the biological responses that may be induced and the severity of any lesions during development. Thus, the vulnerability to EDC exposure is thought to depend on the stage of the development of the organism, where the same dose of an environmental toxicant may have no effect postnatally but can induce malformations in the fetus.

3.5   Effects of Environmental AR Antagonists on Reproductive Development and Androgen Production by Leydig Cells

The environmental anti-androgenic compounds that have been well characterized include procymidone, linuron, vinclozolin and *p,p'*-DDT and its derivatives, all of which are AR antagonists and inhibit androgen-dependent tissue growth *in vivo*.[26] None of these pesticides or their metabolites appears to display significant affinity for the estrogen receptor (ER) or to inhibit 5α-reductase activity *in vitro*.[27] Those compounds compete with endogenous androgens for AR and suppress the expression of certain androgen-dependent target genes.

3.5.1 Vinclozolin

Vinclozolin is a dicarboximide fungicide that has two active metabolites, M1 and M2, which have anti-androgenic properties. These metabolites compete for AR androgen binding and inhibit DHT-induced transcriptional activation by blocking AR binding to androgen response element DNA.[7] Vinclozolin was found to alter the expression of androgen-dependent genes *in vivo* in an anti-androgenic manner.[28] However, vinclozolin and its metabolites have no affinity to the ER but can compete for binding to the progesterone receptor (PR) *in vitro*, although *in vivo* effects associated with the activation of the PR were not demonstrated.[29] The exposure of prepubertal male rats to vinclozolin was shown to affect the hypothalamic–pituitary–gonadal (HPG) axis, inducing elevation in serum LH and testosterone.[30] The vulnerability of the HPG axis to vinclozolin is dependent on the developmental stage of the exposed animals. Stimulation of LH production by the pituitary was not observed in the fetal rat since testis function is not regulated by pituitary hormones during early fetal sex differentiation.[11] Studies *in vivo* have also revealed that exposure to vinclozolin during sexual differentiation demasculinizes and feminizes the male offspring, displaying female-like AGD, retained nipples, hypospadic penis, undescended testes and small sex accessory glands at birth.[31] Further, the most sensitive period of fetal development to anti-androgenic effects of vinclozolin was found to be on E16–E17 in the rat,[32] corresponding to gestational age weeks 15–18 in human fetal development.[6] Experiments *in vitro* have recently revealed that vinclozolin did not affect basal and hCG-stimulated testosterone production by rat Leydig cells in primary culture.[33]

3.5.2 Procymidone

Procymidone [*N*-(3,5-dichlorophenyl)-1,2-dimethylcyclopropane-1,2-dicarbox-imide] is widely used as a fungicide for the control of plant diseases. Procymidone was shown to inhibit DHT-induced AR-DNA binding in a CHO cell promoter interference assay and DHT-induced transcriptional activation in CV-1 cells co-transfected with the human AR and a MMTV-luciferase reporter gene.[34] Similar to vinclozolin, treatment with procymidone on gestational day 14 to postnatal day 3 induced clear anti-androgenic effects on reproductive organ development such as hypospadias, retained nipples, reduced sex accessory gland size and AGD in male pups.[34] It is important to note that, in contrast to developmental effects, procymidone has little effect on the reproductive tract of the adult male rat.[35] Similar to vinclozolin, procymidone as an anti-androgenic compound may inhibit the negative feedback effect of androgen on the hypothalamus and/or the pituitary, thereby resulting in hypergonadotropism. Long-term hypergonadotropism and hyperstimulation of Leydig cells induced by procymidone give rise to interstitial cell tumors in male rats.[36] Furthermore, and accordingly, long-term dietary administration of procymidone to rats has been reported to lead to elevated serum levels of testosterone and LH.[35,36] Recent findings from our own laboratory reveal that dietary administration of procymidone to rats for three months not only enhances serum levels of LH and testosterone, but also that Leydig cells isolated from these animals display an enhanced capacity for producing testosterone in response to stimulation by hCG or Bu_2cAMP, as well as elevated levels of StAR, P450scc and P450c17.[37]

3.5.3 Linuron

Linuron [3-(3,4-dichlorophenyl)-1-methoxy-1-methylurea] is currently being marketed as a selective urea-based herbicide for pre- and/or postemergence control of weeds in crops, including potatoes. This compound binds to both rat prostatic and human AR (hAR) and inhibits DHT-induced gene expression *in vitro*.[38] The effects of linuron treatment on the reproductive system are tightly associated with the developmental stage of the animal exposed, where adult animals are less sensitive to the anti-androgen than fetuses.[39,40] The anti-androgenic effect of linuron is observed clearly upon administration during gestation[40,41] or in a Hershberger assay.[41] For example, administration of linuron (100 mg kg^{-1} daily) to rats during E14–E18 induces hypospadia, decreased AGD and reduced size of androgen-dependent tissues in the male offspring.[40] Similarly, administration of linuron (200 mg kg^{-1} daily) to sexually immature and mature rats for 2 weeks has been shown to decrease the weights of accessory sex organs and increase serum estradiol and LH levels, without altering the serum concentration of testosterone.[38] In contrast to vinclozolin and procymidone, linuron induces a low incidence of hypospadias, but relatively high levels of epididymal and testicular malformations, a profile of changes that is specific for animals exposed *in utero* to phthalates, which suppress fetal Leydig cell steroidogenesis.[40,42] A recent study has shown that linuron treatment *in utero* suppressed *ex vivo* testosterone production, while the expression of steroidogenic acute (StAR) protein and cytochromes P450c17 and P450scc was not affected.[43] However, the fetal testis treated with linuron *in vitro* produced low levels of testosterone with no changes in progesterone production, suggesting that 17β-HSD expression and/or activity is altered by linuron.

Thus, it can be proposed that linuron exerts its anti-androgenic effects by suppressing AR-dependent signaling and inhibiting androgen production by fetal Leydig cells.

3.5.4 DDT Derivatives

The use of DDT as a pesticide has been banned in many countries, but human exposure from DDT can still take place in some developing countries.

p,p'-DDE, a metabolite of the environmentally persistent pesticide DDT, acts as an antagonist of AR both *in vivo* and *in vitro*.[27] When *p,p'*-DDE (100 mg kg^{-1} daily) was administered to rats during development on E14–E18, the AGD was reduced and hypospadia, retention of nipples and reduction in the weights of androgen-dependent tissues occurred.[26] Moreover, long-term administration of *p,p'*-DDE to rat pups from the time of weaning until approximately 50 days of age did not significantly alter the weights of accessory sex organs or serum hormone levels,[27] but did produce pronounced reductions in androgen-dependent tissue weights in the Hershberger assay.[40] As with rats, administration of *p,p'*-DDT or *p,p'*-DDE to rabbits during gestation induces reproductive abnormalities, including cryptorchidism, in the male offspring.[40]

It has also been suggested that *p,p'*-DDE causes abnormalities in the reproductive development of wildlife.[44] For instance, high concentrations of *p,p'*-DDE, together with enhanced susceptibility of the developing reproductive system to this environmental anti-androgen due to a loss of genetic diversity, may be contributing to the high incidence of cryptorchidism in the Florida panther.[44]

Following the ban on DTT, methoxychlor (MC), a derivative of DDT, has found increasing use as an insecticide. Methoxychlor is converted metabolically to several monohydroxy and dihydroxy metabolites that exhibit estrogenic activity.[45] Previous studies have shown that daily exposure of male Long-Evans hooded rats to MC (100 or 200 mg kg^{-1}, by gavage), from the day of weaning to approximately 97–100 days of postnatal age, reduces both circulating levels of testosterone and hCG-stimulated testosterone production by decapsulated testes.[46]

3.6 Anti-androgenic Effects of Phthalate Esters on Reproduction and Leydig Cell Function

Phthalates are widely used as plasticizers in the production of plastics, as well as solvents in inks, and are present, among other places, in food packaging and certain cosmetics.[47] Children on dialysis are suggested to be the most vulnerable group for phthalates exposure, because they may receive high levels of phthalates from dialysis tubing.[48,49] Further, in this context, children and women of childbearing age were demonstrated to have higher levels of phthalates than other groups in general.[24] Moreover, maternal phthalate exposure during pregnancy was reported to induce decreased AGD in male infants,[50] and phthalates present in breast milk are proposed to alter the function of the hypothalamic–pituitary–gonadal axis in male offspring.[51]

Although diethylhexyl phthalate (DEHP) and monoethylhexyl phthalate (MEHP) disturb reproductive development in an anti-androgenic fashion, neither of these compounds binds to the AR or ER.[52,53] It has now been established that these phthalates inhibit Leydig cell steroidogenesis at different stages of fetal development. For example, prenatal exposure of rats to DEHP attenuates serum levels of both LH and testosterone in male offspring at 21 and 35 days of postnatal age.[42] Further, treatment with DEHP for two weeks postnatally (beginning at 21 or 35 days of age) results in a significant decrease in the activity of 17β-HSD and reduces testosterone production by Leydig cells by 50%. On the other hand, treatment of adult rats with DEHP does not influence their Leydig cell steroidogenesis.[42] Thus, it has been concluded that the effects of DEHP on Leydig cell steroidogenesis are dependent on the stage of development at the time of exposure and may involve modulation of the activities of steroidogenic enzymes and/or serum levels of the LH.

In addition to reducing testicular levels of testosterone, inhibition of fetal Leydig cell steroidogenesis by phthalates causes numerous malformations of androgen-dependent tissues in male rats, including a decrease in AGD, hypospadias, epididymal agenesis and testicular atrophy.[20] These abnormalities are suggestive of sub-optimal virilization of the Wolffian duct and urogenital sinus. Furthermore, prenatal exposure of rats to di-*n*-butyl phthalate (DBP) during days 12–21 of gestation gives rise to Leydig cell hyperplasia and adenomas, as well as agenesis of the epididymis. Despite the increased numbers of Leydig cells present after such exposure, testicular levels of testosterone are reduced at E18–E21.[53] However, a recent study has shown that the treatment of pregnant marmosets with MBP (monobutyl phthalate; the active metabolite of dibutyl phthalate) during the masculinization programming window did not alter normal development of the testis at birth or in adulthood in male marmosets.[54] This finding is consistent with the absence of an effect of MBP on steroidogenesis *in vitro* by fetal human testis explants.[55] These findings suggest that the vulnerability to certain types of phthalates can be different in various animal species.

In contrast to other anti-androgenic EDCs, phthalates are able to induce agenesis of the gubernacular cords,[56] whose differentiation is dependent on insl3 produced by Leydig cells.[19,57] This result suggested that phthalates may attenuate insl3 production by fetal Leydig cells, an assumption that was further supported by the demonstration that only phthalates were able to suppress both insl3 and testosterone production by the fetal testis.[58]

Little is currently known about the effects of relatively low doses of phthalates on testicular steroid production by Leydig cells. A recent study has demonstrated that chronic exposures to low environmentally relevant DEHP levels increased plasma levels of LH, testosterone and 17β-estradiol.[59] These metabolic events coincided with high proliferative activity of Leydig cells, likely to be induced by the high levels of LH. Moreover, this study pointed out that DEHP might be indirectly estrogenic because it increased 17β-estradiol plasma levels, presumably due to induction of aromatase activity in Leydig cells by LH.[59] We have recently demonstrated that MEHP inhibits androgen production activated by human chorionic gonadotropin (hCG) in primary cultures of Leydig cells. This process was associated with decreased expression of StAR protein and reduced transport of cholesterol into mitochondria, but no detectable adverse effect on steroidogenic enzymes was observed.[60] Moreover, upon exposure to MEHP alone, 5α-reductase activity was decreased in immature, but not in adult, Leydig cells,[60] indicating higher susceptibility to adverse effects of MEHP in young animals. Recent studies have also demonstrated that phthalates can alter the expression of steroidogenic enzymes involved in the biosynthesis of testosterone. In 2004 it was reported that treatment with DBP attenuates the expression of scavenger receptor B1 (SR-B1), StAR, P450scc, 3β-HSD and P450c17 in the fetal rat testis.[61] Interestingly, the suppression of testosterone synthesis induced by DBP is reversible, with recovery coinciding with the clearance of DBP and its testicular metabolite.

3.7 Effects of Dioxins on Development of Reproductive Organs and Leydig Cell Function

Polychlorinated dibenzo-*p*-dioxins, such as 2,3,7,8-tetrachlorodibenzo-*p*-dioxin (TCDD), are released into the environment as a byproduct of the manufacture of chlorinated hydrocarbons and other industrial processes, and are recognized as being highly potent developmental and reproductive toxins. The toxicity of TCDD is mediated by the aryl hydrocarbon receptor (AhR). Upon binding of TCDD, the receptor complex translocates into the nucleus, where AhR heterodimerizes with the aryl hydrocarbon receptor nuclear translocator (ARNT), binds to dioxin-responsive enhancer elements and regulates the transcription of target genes.[62] Both AhR and ARNT are expressed in the adult male rat reproductive tract (testis, epididymis, vas deference, seminal vesicle, prostate).[63]

The sensitivity of the male rat reproductive system to TCDD is greatly influenced by the timing of exposure.[64] Deleterious effects of TCDD on the development and function are largely determined by its suppressive influence on Leydig cell function. It is generally accepted that TCDD suppresses steroidogenesis in Leydig cells by inhibiting the expression and/or activity of several key steroidogenic enzymes.[65–67] TCDD treatment was shown to inhibit the mobilization of cholesterol and of the activities of cytochromes P450scc and P450c17 in Leydig cells.[65,66] Many or all of these alterations in steroidogenesis might be explained by the recent finding that TCDD inhibits the formation of cAMP by cultured Leydig cells.[67]

Moreover, this toxicant can induce several morphological changes in Leydig cells associated with reduced size and decreased volume of the smooth endoplasmic reticulum.[68] Additional experiments have demonstrated that pharmacological treatment of rats with hCG preserves the Leydig cell cytoplasm and organelles in sufficient quantities to prevent this disturbance of Leydig cell function by TCDD.[69]

Interestingly, a single dose of TCDD (ranging from 50 ng to 2 μg kg^{-1}) to rats or hamsters during sexual differentiation gives rise to a number of reproductive alterations in the male, including delayed puberty and reductions in the size of the ventral prostate, seminal vesicle and testis.[70] The observation that these animals do not display decreased AGD or areolas suggests that TCDD may also be affecting the developing reproductive tract *via* pathways that do not involve the AR. Further, a recent study has shown that treatment of pregnant rats with TCDD on E11 reduced intratesticular testosterone and plasma LH levels in fetal males on E19.5.[71]

Much less is known about the potential effects of TCDD on the function of human Leydig cells. The report that serum levels of dioxin correlated positively with serum levels of LH, and negatively with testosterone levels, in men exposed to dioxin in a chemical plant indicates that TCDD may exert a direct toxic effect on the human Leydig cell.[72]

3.8 Effects of Estrogenic Compounds on Fertility and Hormonal Function of Leydig Cells

It is generally accepted that estrogens exert a negative effect on testicular steroidogenesis and reproductive tract development in different classes of animals.[6,73] ERα is expressed by rodent Leydig cells[74] and ERα null mice were shown to have lower sperm counts and elevated serum testosterone levels.[75] Moreover, the expression of several steroidogenic enzymes and factors such as P450c17, 17β-HSD and StAR was higher in ERα-deleted mice than in wild-type mice,[76] suggesting that ERα signaling can suppress androgen production in Leydig cells. It is well known that exposure to diethylstilbestrol (DES) by pregnant women results in increased occurrence of hypospadias in their sons.[77] Similarly, hypospadias and other genital abnormalities have been reported in boys whose mothers were exposed to 17α-ethynylestradiol in pregnancy.[78] These effects of the estrogenic compounds can be due to suppression of fetal Leydig cell steroidogenesis and/or a direct effect on the developing male external genitalia.[6] However, the finding that fetal human Leydig cells do not express ERα,[79] the main ER executing estrogenic effects in rodents, suggest that environmental estrogen compounds may exert their effects *via* ERβ, which is expressed by fetal human Leydig cells.[79] Studies on rodents exposed to DES during the neonatal period have shown that estrogens not only inhibit fetal Leydig cell testosterone production but also induce a profound loss of AR protein expression in the reproductive organs.[73,80] Thus, estrogens could affect the development of AR-dependent organs, irrespective of any effect on testosterone production by the fetal testis.[6]

Since exogenous compounds of anthropogenic or natural origin with estrogenic functions are widely distributed in food, air and water, it is suggested that such environmental factors may negatively influence reproductive functions in humans. Phytoestrogens are a various group of naturally occurring plant compounds. They are structurally similar to 17β-estradiol and therefore have the ability to cause estrogenic or/and anti-estrogenic effects.[81] Chemically, the phytoestrogens can be divided into four main classes: flavonoids (*e.g.* genistein, daidazen, naringenin and kaempferol), coumestans (coumestrol), lignans (matairesinol and secoilariciresinol) and stilbenes (resveratrol). The main sources of phytoestrogens are fruits, vegetables and legumes.[82] It is important to note that humans may ingest high milligram quantities of flavanoids (*e.g.* genistein, daidazen) *via* food consumption. Low micromolar levels of genistein are found in the serum of the Japanese population[83] and in infants who consume large amounts of food products derived from soybeans.[84] The isoflavones genistein and daidzein are widely distributed in human and animal diets and were found to have a structural similarity to 17β-estradiol.[85] The highest amount of flavonoids has been found in soybeans and soy food (1.2–3.3 mg isoflavones g^{-1} dry weight) as well as in other types of beans and legumes.[82] The presence of genistein and daidzein in soy-based infant formula at high micromolar concentrations, and the fact that blood concentrations of the iso-flavones are 1000 times higher than endogenous estradiol,[84] indicate a possibility for interaction with Leydig cell function and the development of the reproductive system in male infants. Lower concentrations of serum testosterone associated with an increased number of Leydig cells have been observed in neonatal marmosets fed with soy milk formula when compared with animals fed with cow's milk formula.[86] We have demonstrated recently that long-term dietary administration of genistein to rats suppressed the steroidogenic response of Leydig cells to hCG and Bu_2cAMP by down-regulating the expression of P450scc,[37] suggesting that genistein and/or its metabolites have direct effects on Leydig cell function. The exact cellular and molecular mechanisms behind these effects of genistein remain to be defined, but may involve activation of the ER by the phytoestrogen. Further, a recent study has demonstrated that genistein may suppress steroidogenesis in Leydig cells *in vitro* by affecting the adenylate cyclase function *via* uncoupling of LHR from G proteins.[87]

Recently, we have demonstrated that the prenyl flavonoids 8-prenyl-naringenin and isoxanthohumol, the phytoestrogens found in the hop plant, *Humulus lupulus* (Cannabaceae), which is traditionally used to add bitterness and flavor to beer,[88] exert complex maturation-dependent effects on Leydig cell steroidogenesis. Those compounds inhibited hCG-stimulated androgen production by Leydig cells at all stages of their development, a process that was associated with reduced ability of the cells to produce cAMP.[89] Further, in this context, we have reported that the phytoestrogen resveratrol and certain resveratrol analogs structure-dependently attenuated steroidogenesis in Leydig cells through suppression of the expression of StAR and cytochrome P450c17.[90]

Another synthetic compound with estrogenic activity, which is widely used in industry, is bisphenol A (BPA). It stimulates estrogen-dependent cellular proliferation and induces the expression of progesterone receptors.[91] The chemical structure of BPA is quite similar to that of the potent estrogenic compound DES but it is a much weaker estrogen, approximately 1000- to 2000-fold less potent than estradiol.[91] It was demonstrated that exposure to environmentally relevant BPA levels has adverse effects on testicular function by decreasing pituitary LH secretion and reducing the expression of P450c17 and testosterone production in Leydig cells *ex vivo*.[92] BPA was found to act directly on Leydig cells because it decreased testosterone production after treatment of Leydig cells *in vitro*. Inhibition of steroidogenesis was ER mediated and associated with inhibition of enzyme activity.[92] Analysis of the expression of steroidogenic enzymes by RT-PCR indicated that BPA may cause specific inhibition of the P450c17 enzyme, which is known to be susceptible to inhibition by estradiol.[76]

3.9 Concluding Remarks

EDCs of natural or anthropogenic origin may cause deleterious effects on male reproductive function. An overview of harmful effects of EDCs on development of the reproductive organs is presented in Table 3.1. The mechanisms of their adverse actions may be diverse but one important end point is reduced capacity of Leydig cells to produce androgens. It is important to note that there are species-dependent differences in the activation of fetal Leydig cell steroidogenesis between humans (gonadotropin dependent) and rodents (paracrine factors dependent) during the period of masculinization of the reproductive tract. Moreover, androgen production by fetal Leydig cells is different in humans and rodents (the Δ^5 pathway *vs.* the Δ^4 pathway, respectively). These differences may determine a different vulnerability of fetal Leydig cells of humans and rodents to EDCs.

Other discrepancies that need to be resolved are the capacity of phthalates and their metabolites to suppress androgen production in human fetal Leydig cells. Experiments on human fetal testis explants *in vitro* have disclosed no

Table 3.1 Natural and synthetic chemicals affecting development of male reproductive organs.

Substance	Target	Malformation/adverse action	Application or source
Procymidone	AR (antagonist)	Retained nipples, hypospadia, agenesis of ventral prostate, undescended testes, hypergonadotropism	Fungicide; control of plant diseases
Linuron	AR (antagonist)	Epididymal and testicular malformations, suppression of fetal Leydig cell steroidogenesis	Herbicide; post emergence control of weeds in crops
Vinclozolin	AR (antagonist)	Similar to procymidone	Fungicide
p,p'-DDT	AR (antagonist)	Cryptorchidism, decreased weights of accessory sex organs	Pesticide
Dioxins	Aryl hydrocarbon receptor (agonist)	Toxic effect on Leydig cells, suppression of the development of the ventral prostate, seminal vesicles and testis	Byproduct of chlorinated hydrocarbons
Phthalates	Peroxisome proliferator-activated receptors (PPARs) (agonist)	Epididymal and testis abnormalities, hypospadia, inhibition of androgen production by Leydig cells	Plasticizers
Genistein	ERs (agonist)	Suppression of Leydig cell steroidogenesis	Soy-derived food
Resveratrol	ERs (agonist)	Similar to genistein	Red wine, red grapes
Bisphenol A	ERs (agonist)	Suppression of LH production and Leydig cell steroidogenesis	Synthesis of polycarbonate plastics

effects of MBP/MEHP on steroidogenesis and the conclusion on the negative effects of phthalates on the development of the reproductive tract in humans is mostly based on associations between maternal phthalate exposure and lower AGD in the resulting sons.

Dysfunction of fetal Leydig cells induced by EDCs may cause incomplete masculinization and development of various malformations in the male reproductive tract of humans and animals.

Attenuation of adult Leydig cell development and function may manifest as suppressed spermatogenesis and libido. The impaired Leydig cell function is displayed by a decrease in testosterone production as a consequence of reduced expression of several important steroidogenic factors, including StAR, P450c17, P450scc and 3β-HSD. All these factors are important components of the steroidogenic machinery and may serve as susceptible targets of EDC actions (Figure 3.1). However, our knowledge on the intracellular signaling cascades triggered by EDCs is still limited. Little is also known about the mechanisms that determine the selectivity of different EDCs for disruption of particular target molecules involved in steroidogenesis. Increasing evidence indicates that multiple mechanisms contribute to endocrine disruption, including: (1) direct effects on the expression of genes of steroidogenic factors, (2) suppression of upstream signaling pathways controlling phosphorylation

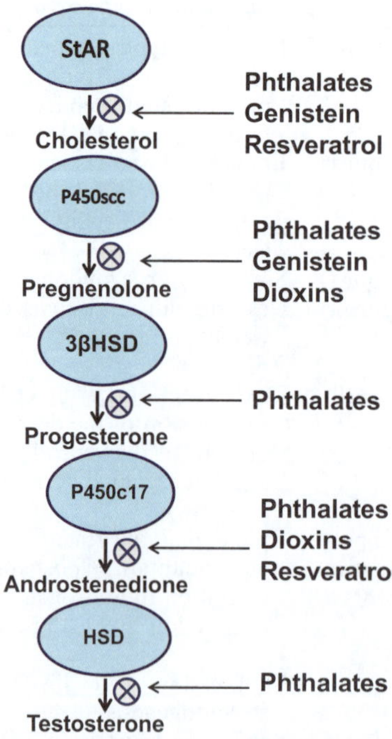

Figure 3.1 Proposed sites of action of EDCs on the steroidogenic machinery of Leydig cells. Suppression of the enzyme expression and/or activity is indicated by ⊗.

and activation of target proteins involved in transport of cholesterol into mitochondria of steroidogenic cells, (3) direct interference with AR function blocking androgen action, and (4) epigenetic modifications resulting in changes in gene expression profiles. The action of EDCs on Leydig cell function and the reproductive potential is a complex process that depends on the exposure route, dose, the developmental stage of the exposed target organism and many other factors. Together, these factors determine the potential risk for adverse consequences with long-lasting effects on male reproductive function.

Acknowledgements

The authors own work cited was supported financially by grants from the Swedish Research Council (no. 8282), the Swedish Children's Cancer Fund, the Frimurare Barnhuset Foundation in Stockholm, Sällskapet Barnavård, Stiftelsen Samariten and the Karolinska Institute.

References

1. N. E. Skakkebaek, E. Rajpert-De Meyts and K. M. Main, *Hum. Reprod.*, 2001, **16**, 972.
2. K. A. Boisen, M. Kaleva, K. M. Main, H. E. Virtanen, A. M. Haavisto, I. M. Schmidt, M. Chellakooty, I. N. Damgaard, C. Mau, M. Reunanen, N. E. Skakkebaek and J. Toppari, *Lancet*, 2004, **363**, 1264.
3. M. Welsh, P. T. Saunders, M. Fisken, H. M. Scott, G. R. Hutchison, L. B. Smith and R. M. Sharpe, *J. Clin. Invest.*, 2008, **118**, 1479.
4. H. M. Scott, G. R. Hutchison, M. S. Jobling, C. McKinnell, A. J. Drake and R. M. Sharpe, *Endocrinology*, 2008, **149**, 5280.
5. J. D. Wilson, *Annu. Rev. Physiol.*, 1978, **40**, 279.
6. H. M. Scott, J. I. Mason and R. M. Sharpe, *Endocrin. Rev.*, 2009, **30**, 883.
7. C. Wong, W. R. Kelce, M. Sar and E. M. Wilson, *J. Biol. Chem.*, 1995, **270**, 19998.
8. J. Tan, K. B. Marschke, K. C. Ho, S. T. Perry, E. M. Wilson and F. S. French, *J. Biol. Chem.*, 1992, **267**, 7958.
9. G. R. Brown, C. M. Nevison, H. M. Fraser and A. F. Dixson, *Int. J. Androl.*, 1999, **22**, 119.
10. L. J. Pelliniemi and M. Niei, *Z. Zellforsch. Mikrosk. Anat.*, 1969, **99**, 507.
11. R. Habert, H. Lejeune and J. M. Saez, *Mol. Cell. Endocrinol.*, 2001, **179**, 47.
12. P. J. O'shaughnessy, P. J. Baker and H. Johnston, *Int. J. Androl.*, 2006, **29**, 90.
13. R. L. Molsberry, B. R. Carr, C. R. Mendelson and E. R. Simpson, *J. Clin. Endocrinol. Metab.*, 1982, **55**, 791.
14. R. Lambrot, H. Coffigny, C. Pairault, A. C. Donnadieu, R. Frydman, R. Habert and V. Rouiller-Fabre, *J. Clin. Endocrinol. Metab.*, 2006, **91**, 2696.
15. F. E. Carr and W. W. Chin, *Endocrinology*, 1985, **116**, 1151.
16. M. L. Aubert, M. Begeot, B. P. Winiger, G. Morel, P. C. Sizonenko and P. M. Dubois, *Endocrinology*, 1985, **116**, 1565.

17. X. Ma, Y. Dong, M. M. Matzuk and T. R. Kumar, *Proc. Natl. Acad. Sci. USA*, 2004, **101**, 17294.
18. A. H. Payne and D. B. Hales, *Endocrin. Rev.*, 2004, **25**, 947.
19. R. Ivell, M. Balvers, R. Domagalski, H. Ungefroren, N. Hunt and W. Schulze, *Mol. Hum. Reprod.*, 1997, **3**, 459.
20. L. E. Gray, Jr., J. Ostby, J. Furr, C. J. Wolf, C. Lambright, L. Parks, D. N. Veeramachaneni, V. Wilson, M. Price, A. Hotchkiss, E. Orlando and L. Guillette, *Hum. Reprod. Update*, 2001, **7**, 248.
21. U. N. Joensen, N. Jorgensen, E. Rajpert-De Meyts and N. E. Skakkebaek, *Basic Clin. Pharmacol. Toxicol.*, 2008, **102**, 155.
22. D. A. Murature, S. Y. Tang, G. Steinhardt and R. C. Dougherty, *Biomed. Environ. Mass Spectrom.*, 1987, **14**, 473.
23. A. Giwercman, E. Carlsen, N. Keiding and N. E. Skakkebaek, *Environ. Health Perspect.*, 1993, **101**(suppl. 2), 65.
24. B. C. Blount, M. J. Silva, S. P. Caudill, L. L. Needham, J. L. Pirkle, E. J. Sampson, G. W. Lucier, R. J. Jackson and J. W. Brock, *Environ. Health Perspect.*, 2000, **108**, 979.
25. W. Foster, S. Chan, L. Platt and C. Hughes, *J. Clin. Endocrinol. Metab.*, 2000, **85**, 2954.
26. L. E. Gray, Jr., J. Ostby, E. Monosson and W. R. Kelce, *Toxicol. Ind. Health*, 1999, **15**, 48.
27. W. R. Kelce, C. R. Stone, S. C. Laws, L. E. Gray, Jr., J. A Kemppainen and E. M. Wilson, *Nature*, 1995, **375**, 581.
28. W. R. Kelce, C. R. Lambright, L. E. Gray, Jr. and K. P. Roberts, *Toxicol. Appl. Pharmacol.*, 1997, **142**, 192.
29. S. C. Laws, S. A. Carey, W. R. Kelce, R. L. Cooper and L. E. Gray, Jr., *Toxicology*, 1996, **112**, 173.
30. E. Monosson, W. R. Kelce, C. Lambright, J. Ostby and L. E. Gray, Jr., *Toxicol. Ind. Health*, 1999, **15**, 65.
31. J. Hellwig, B. van Ravenzwaay, M. Mayer and C. Gembardt, *Regul. Toxicol. Pharmacol.*, 2000, **32**, 42.
32. C. J. Wolf, G. A. LeBlanc, J. S. Ostby and L. E. Gray, Jr., *Toxicol. Sci.*, 2000, **55**, 152.
33. E. P. Murono and R. C. Derk, *Reprod. Toxicol.*, 2004, **19**, 135.
34. J. Ostby, W. R. Kelce, C. Lambright, C. J. Wolf, P. Mann and L. E. Gray, Jr., *Toxicol. Ind. Health*, 1999, **15**, 80.
35. S. Hosokawa, M. Murakami, M. Ineyama, T. Yamada, Y. Koyama, Y. Okuno, A. Yoshitake, H. Yamada and J. Miyamoto, *J. Toxicol. Sci.*, 1993, **18**, 111.
36. M. Murakami, S. Hosokawa, T. Yamada, M. Harakawa, M. Ito, Y. Koyama, J. Kimura, A. Yoshitake and H. Yamada, *Toxicol. Appl. Pharmacol.*, 1995, **131**, 244.
37. K. Svechnikov, V. Supornsilchai, M. L. Strand, A. Wahlgren, D. Seidlova-Wuttke, W. Wuttke and O. Soder, *J. Endocrinol.*, 2005, **187**, 117.
38. J. C. Cook, L. S. Mullin, S. R. Frame and L. B. Biegel, *Toxicol. Appl. Pharmacol.*, 1993, **119**, 195.

39. B. S. McIntyre, N. J. Barlow, M. Sar, D. G. Wallace and P. M. Foster, *Reprod. Toxicol.*, 2002, **16**, 131.

40. L. E. Gray, Jr., C. Wolf, C. Lambright, P. Mann, M. Price, R. L. Cooper and J. Ostby, *Toxicol. Ind. Health*, 1999, **15**, 94.

41. C. Lambright, J. Ostby, K. Bobseine, V. Wilson, A. K. Hotchkiss, P. C. Mann and L. E. Gray, Jr., *Toxicol. Sci.*, 2000, **56**, 389.

42. B. T. Akingbemi, R. T. Youker, C. M. Sottas, R. Ge, E. Katz, G. R. Klinefelter, B. R. Zirkin and M. P. Hardy, *Biol. Reprod.*, 2001, **65**, 1252.

43. V. S. Wilson, C. R. Lambright, J. R. Furr, K. L. Howdeshell and L. E. Gray, Jr., *Toxicol. Lett.*, 2009, **186**, 73.

44. C. F. Facemire, T. S. Gross and L. J. Guillette, Jr., *Environ. Health Perspect.*, 1995, **103**(suppl. 4), 79.

45. W. H. Bulger, R. M. Muccitelli and D. Kupfer, *Biochem. Pharmacol.*, 1978, **27**, 2417.

46. L. E. Gray, Jr., J. Ostby, J. Ferrell, G. Rehnberg, R. Linder, R. Cooper, J. Goldman, V. Slott and J. Laskey, *Fundam. Appl. Toxicol.*, 1989, **12**, 92.

47. J. A. Thomas and M. J. Thomas, *J. Crit. Rev. Toxicol.*, 1984, **13**, 283.

48. J. W. Brock, S. P. Caudill, M. J. Silva, L. L. Needham and E. D. Hilborn, *Bull. Environ. Contam. Toxicol.*, 2002, **68**, 309.

49. R. Hauser, S. Duty, L. Godfrey-Bailey and A. M. Calafat, *Environ. Health Perspect.*, 2004, **112**, 751.

50. S. H. Swan, K. M. Main, F. Liu, S. L. Stewart, R. L. Kruse, A. M. Calafat, C. S. Mao, J. B. Redmon, C. L. Ternand, S. Sullivan and J. L. Teague, *Environ. Health Perspect.*, 2005, **113**, 1056.

51. K. M. Main, G. K. Mortensen, M. M. Kaleva, K. A. Boisen, I. N. Damgaard, M. Chellakooty, I. M. Schmidt, A. M. Suomi, H. E. Virtanen, D. V. Petersen, A. M. Andersson, J. Toppari and N. E. Skakkebaek, *Environ. Health Perspect.*, 2006, **114**, 270.

52. L. G. Parks, J. S. Ostby, C. R. Lambright, B. D. Abbott, G. R. Klinefelter, N. J. Barlow and L. E. Gray, Jr., *Toxicol. Sci.*, 2000, **58**, 339.

53. E. Mylchreest, M. Sar, D. G. Wallace and P. M. Foster, *Reprod. Toxicol.*, 2002, **16**, 19.

54. C. McKinnell, R. T. Mitchell, M. Walker, K. Morris, C. J. Kelnar, W. H. Wallace and R. M. Sharpe, *Hum. Reprod.*, 2009, **24**, 2244.

55. N. Hallmark, M. Walker, C. McKinnell, I. K. Mahood, H. Scott, R. Bayne, S. Coutts, R. A. Anderson, I. Greig, K. Morris and R. M. Sharpe, *Environ. Health. Perspect.*, 2007, **115**, 390.

56. L. E. Gray, Jr., J. Ostby, J. Furr, M. Price, D. N. Veeramachaneni and L. Parks, *Toxicol. Sci.*, 2000, **58**, 350.

57. S. Nef and L. F. Parada, *Nat. Genet.*, 1999, **22**, 295.

58. V. S. Wilson, C. Lambright, J. Furr, J. Ostby, C. Wood, G. Held and L. E. Gray, Jr., *Toxicol. Lett.*, 2004, **146**, 207.

59. B. T. Akingbemi, R. Ge, G. R. Klinefelter, B. R. Zirkin and M. P. Hardy, *Proc. Natl. Acad. Sci. USA*, 2004, **101**, 775.

60. K. Svechnikov, I. Svechnikova and O. Soder, *Reprod. Toxicol.*, 2008, **25**, 485.

61. C. J. Thompson, S. M. Ross and K. W. Gaido, *Endocrinology*, 2004, **145**, 1227.
62. J. Mimura and Y. Fujii-Kuriyama, *Biochim. Biophys. Acta*, 2003, **1619**, 263.
63. B. L. Roman, R. S. Pollenz and R. E. Peterson, *Toxicol. Appl. Pharmacol.*, 1998, **150**, 228.
64. D. L. Bjerke and R. E. Peterson, *Toxicol. Appl. Pharmacol.*, 1994, **127**, 241.
65. C. A. Mebus, V. R. Reddy and W. N. Piper, *Biochem. Pharmacol.*, 1987, **36**, 727.
66. R. W. Moore, C. R. Jefcoate and R. E. Peterson, *Toxicol. Appl. Pharmacol.*, 1991, **109**, 85.
67. K. P. Lai, M. H. Wong and C. K. Wong, *J. Endocrinol.*, 2005, **185**, 519.
68. L. Johnson, C. E. Wilker, S. H. Safe, B. Scott, D. D. Dean and P. H. White, *Toxicology*, 1994, **89**, 49.
69. C. E. Wilker, T. H. Welsh, Jr., S. H. Safe, T. R. Narasimhan and L. Johnson, *Toxicology*, 1995, **95**, 93.
70. L. E. Gray, Jr., W. R. Kelce, E. Monosson, J. S. Ostby and L. S. Birnbaum, *Toxicol. Appl. Pharmacol.*, 1995, **131**, 108.
71. A. Adamsson, U. Simanainen, M. Viluksela, J. Paranko and J. Toppari, *Int. J. Androl.*, 2009, **32**, 575.
72. G. M. Egeland, M. H. Sweeney, M. A. Fingerhut, K. K. Wille, T. M. Schnorr and W. E. Halperin, *Am. J. Epidemiol.*, 1994, **139**, 272.
73. A. Rivas, J. S. Fisher, C. McKinnell, N. Atanassova and R. M. Sharpe, *Endocrinology*, 2002, **143**, 4797.
74. G. Pelletier, C. Labrie and F. J. Labrie, *J. Endocrinol.*, 2000, **165**, 359.
75. E. M. Eddy, T. F. Washburn, D. O. Bunch, E. H. Goulding, B. C. Gladen, D. B. Lubahn and K. S. Korach, *Endocrinology*, 1996, **137**, 4796.
76. B. T. Akingbemi, R. Ge, C. S. Rosenfeld, L. G. Newton, D. O. Hardy, J. F. Catterall, D. B. Lubahn, K. S. Korach and M. P. Hardy, *Endocrinology*, 2003, **144**, 84.
77. H. Klip, J. Verloop, J. D. van Gool, M. E. Koster, C. W. Burger and F. E. van Leeuwen, *Lancet*, 2002, **359**, 1102.
78. E. D. Whitehead and E. Leiter, *J. Urol.*, 1981, **125**, 47.
79. T. L. Gaskell, L. L. Robinson, N. P. Groome, R. A. Anderson and P. T. Saunders, *J. Clin. Endocrinol. Metab.*, 2003, **88**, 424.
80. C. McKinnell, N. Atanassova, K. Williams, J. S. Fisher, M. Walker, K. J. Turner, T. K. Saunders and R. M. Sharpe, *J. Androl.*, 2001, **22**, 323.
81. G. G. Kuiper, J. G. Lemmen, B. Carlsson, J. C. Corton, S. H. Safe, P. T. van der Saag, B. van der Burg and J. A. Gustafsson, *Endocrinology*, 1998, **139**, 4252.
82. K. Reinli and G. Block, *Nutr. Cancer*, 1996, **26**, 123.
83. H. Adlercreutz, H. Markkanen and S. Watanabe, *Lancet*, 1993, **342**, 1209.
84. K. D. Setchell, L. Zimmer-Nechemias, J. Cai and J. E. Heubi, *Lancet*, 1997, **350**, 23.
85. B. T. Akingbemi, T. D. Braden, B. W. Kemppainen, K. D. Hancock, J. D. Sherrill, S. J. Cook, X. He and J. Supko, *Endocrinology*, 2007, **148**, 4475.

86. R. M. Sharpe, B. Martin, K. Morris, I. Greig, C. McKinnell, A. S. McNeilly and M. Walker, *Hum. Reprod.*, 2002, **17**, 1692.
87. K. D. Hancock, E. S. Coleman, Y. X. Tao, E. E. Morrison, T. D. Braden, B. W. Kemppainen and B. T. Akingbemi, *Toxicol. Lett.*, 2009, **184**, 169.
88. J. F. Stevens and J. E. Page, *Phytochemistry*, 2004, **65**, 1317.
89. G. Izzo,O. Söder and K. Svechnikov, *J. Appl. Toxicol.*, 2010, in press.
90. K. Svechnikov, C. Spatafora, I. Svechnikova, C. Tringali and O. Soder, *J. Appl. Toxicol.*, 2009, **29**, 673.
91. A. V. Krishnan, P. Stathis, S. F. Permuth, L. Tokes and D. Feldman, *Endocrinology*, 1993, **132**, 2279.
92. B. T. Akingbemi, C. M. Sottas, A. I. Koulova, G. R. Klinefelter and M. P. Hardy, *Endocrinology*, 2004, **145**, 592.

Marked For Life: How Environmental Factors Affect the Epigenome

PAULIINA DAMDIMOPOULOU,[b] STEFAN WEIS,[a] IVAN NALVARTE[c] AND JOËLLE RÜEGG[a]

[a] Department of Biomedicine, Institute of Biochemistry and Genetics, University of Basel, Mattenstrasse 28, 4058 Basel, Switzerland; [b] Department of Biosciences and Nutrition, Karolinska Institute, Hälsovägen 7, 14152 Huddinge, Sweden; [c] Friedrich Miescher Institute, Maulbeerstrasse 66, 4058 Basel, Switzerland

4.1 Introduction

The risk of diseases varies globally across different geographical locations. In particular, the prevalence of different types of cancers is distinct between industrialized and developing countries. When individuals migrate from one region to another, the risk levels adjust to the local level within a few genera-tions.[1] This indicates that the susceptibility to develop non-communicable diseases, like cardiovascular disease and cancer, depends on an interplay between the genes and the environment, instead of genes only. Interestingly, early development seems to be a particularly sensitive time period for environmental influences. Extrinsic factors, such as maternal stress hormones, dietary constituents and endocrine disruptive chemicals (EDCs), are known to affect fetal development and later disease phenotypes in experimental animals. Phenotypes that are induced by developmental exposures are often attributed to epigenetic changes, i.e. inheritable changes in gene function that are not determined by the

Issues in Toxicology No. 11
Hormone-Disruptive Chemical Contaminants in Food
Edited by Ingemar Pongratz and Linda Vikström Bergander

Published by the Royal Society of Chemistry, www.rsc.org

DNA sequence. Indeed, many different environmental factors have been shown to modify the epigenetic landscape of developing experimental animals, with a consequent change in the adult phenotype. Whether fetal exposure to adverse conditions leads to the epigenetic changes underlying non-communicable diseases in humans is an intensively studied question.

In this chapter, we will discuss the possible involvement of environmental factors, in particular EDCs, in shaping human health and risk of disease through epigenetic mechanisms. We start by describing different mechanisms to code epigenetic information and then move on to discussing how early life influences can affect this coding and what kind of consequences it can have on the level of the phenotype. Finally, we will summarize the mechanistic data on the effects of endocrine disruptors on the epigenetic machinery.

4.2   The Epigenome

The primary DNA sequence has long been recognized to contain the "blueprint" of life, *i.e.* the genes. The orchestrated activity of genes is fundamental in the ontogenesis of all living organisms. Hence, gene activity must be regulated in response to intrinsic and extrinsic signals. Mechanisms underlying the clonally heritable regulation of gene activity form a second layer of genomic information referred to as "epigenetic". More than 70 years ago the term "epigenetics" was used for the first time, and today it is defined as "the study of mitotically and/or meiotically heritable changes in gene function that cannot be explained by changes in DNA sequence".[2] Epigenetic information exists as chemical modifications to nucleotide bases (DNA methylation) and histone proteins. Furthermore, small regulatory RNA molecules play a role in epigenetic control. In this chapter we focus on DNA methylation and histone modifications.

4.2.1   DNA Methylation

DNA methylation is a covalent chemical modification of nucleic acid bases in DNA at the C5 position of a cytosine residue (Figure 4.1A). Originally, cytosine methylation was discovered in bacteria, but was soon after also reported in the DNA of animals and plants.[3,4] In animals, DNA methylation occurs almost exclusively in a CpG context, where the modified cytosine is $3'$-flanked by a guanine. In humans, about 70% of these CpG dinucleotides appear methylated. The majority of methylated CpGs are associated with silenced and condensed chromatin and resides within repetitive DNA elements in heterochromatin.[5] A small fraction of all CpG sites (about 7% in humans) appears in clusters, so-called CpG islands. More than half of the human gene promoters are associated with such CpG islands and their methylation status is linked to gene expression.[6–8] Promoter CpG island methylation is generally associated with gene silencing, whereas unmethylated promoter CpG islands most frequently correlate with active transcription (Figure 4.1B).[9]

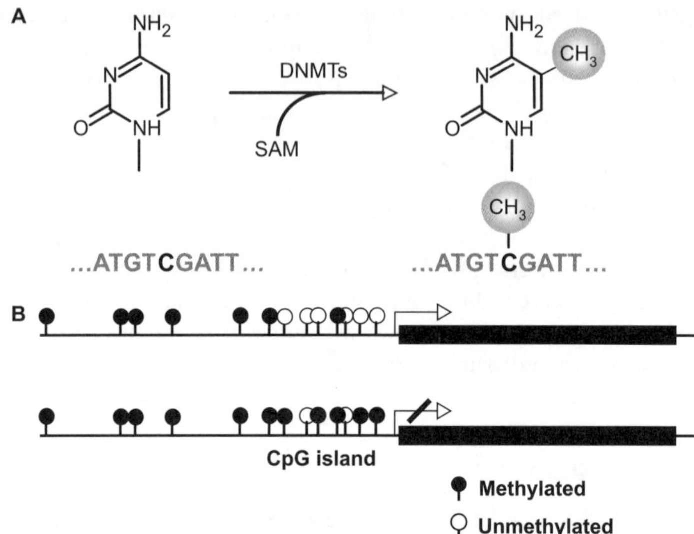

Figure 4.1 DNA methylation. (A) Methylation of cytosine at the C5 position is enzymatically catalyzed by DNA methyltransferases (DNMTs). The methylation reaction is reversible *via* active and passive mechanisms. Mammalian DNA methylation occurs almost exclusively in the CpG sequence context. (B) CpG sites [depicted by *open* (unmethylated) and *filled* (methylated) lollipops] are distributed over the whole mammalian genome. They occur clustered in CpG islands in the promoter region of about 60% of the human genes, and their methylation associates with transcriptional repression of the respective gene. CpG sites located in intergenic regions are mostly methylated and often associated with repetitive, inactive DNA elements. SAM, *S*-adenosylmethionine.

Cytosine methylation is catalyzed by DNA methyltransferases (DNMTs). These enzymes use *S*-adenosylmethionine (SAM) as a methyl donor and transfer it to the cytidyl moiety (Figure 4.1A). Among the five identified human DNA methyltransferases, only DNMT1, DNMT3A and DNMT3B show methyltransferase activity.[10] DNMT1 localizes to DNA replication foci and copies the methylation pattern to the newly synthesized unmethylated DNA strand. DNMT1 provides the enzymatic basis of epigenetic inheritance and is therefore also referred to as the maintenance DNA methyltransferase. DNMT3A and DNMT3B are directed to target DNA sequences devoid of preexisting methylation and catalyze the methylation of individual CpG sites. They therefore constitute the *de novo* DNA methyltransferases.

In contrast to the detailed knowledge about establishing the cytosine methylation mark, the reverse process of DNA demethylation is poorly understood. Because of the strength of the C–C bond, methylation of cytosine was regarded as irreversible. It was believed that passive, replication-dependent demethylation, *i.e.* inhibition of the maintenance DNA methyltransferase during replication, is the only way of DNA demethylation. However, studies in mouse embryos have revealed that methylation can be removed independently of DNA replication by an active mechanism. Based on various findings, different mechanisms of active DNA demethylation have been suggested, many of them ascribing DNA repair pathways a central role.[11-13]

In all higher eukaryotes, DNA methylation constitutes an essential mechanism for dynamic chromatin organization and maintenance of genomic stability. It is involved in the regulation of chromatin structure and transcriptional activity, and dynamic DNA methylation changes have an essential part in normal embryogenesis and cell differentiation.[14,15] During early embryogenesis in mouse it has been observed that DNA methylation marks in the paternal pronucleus are erased and the maternal genome becomes passively demethylated already before the egg is implanted in the uterus (Figure 4.2A).[16] Genome-wide DNA methylation levels increase rapidly in the blastocyst, establishing methylation patterns that already differ between the inner cell mass and the outer cell layer. Further cell differentiation is accompanied by an increasing specification of promoter methylation patterns, reflecting the emerging tissue-specific gene expression patterns. Thereby, pluropotency genes (genes that are necessary to keep stem cells in an undifferentiated state) are silenced whereas tissue-specific gene promoters are selectively activated.[17,18] Cells in the germ line (primordial germ cells) undergo a second genome-wide demethylation before DNA methylation marks are re-established in a gender-specific manner (Figure 4.2B).[16] Imprinted genes have an exceptional role in this cycle of methylation and demethylation. The methylation status, and hence their expression, depends only on their parental origin, *i.e.* imprinted genes are mono-allelically expressed either from the paternal or the maternal allel. They resist the genome-wide demethylation at the preimplantation stage as well as the global *de novo* methylation during implantation. Imprinted genes are implicated in regulation of fetal growth and postnatal behavior. Mammalian genomic imprinting is proposed to represent a mechanism by which the gender-specific conflicting interests in fetal growth are expressed.[16]

Since DNA methylation is involved in numerous cellular functions, aberrant DNA methylation often associates with human pathologies, including auto-immunity disorders (systemic lupus erythematosus), developmental diseases (fragile X and Prader–Willi syndromes) and cancer. In many cancers a global change of methylation pattern is observed, with a loss of methylation in repetitive heterochromatic regions leading to chromosomal instability, and promoter hypermethylation leading to aberrant gene silencing of, for example, tumor suppressor genes.[19,20]

4.2.2 Histone Modifications

Chromatin is the condensed proteinous structure in which DNA is packed in the cell nuclei. The fundamental unit of chromatin is the nucleosome, which is composed of an octamer of the four core histone proteins (H3, H4, H2A and H2B), around which 147 base pairs of DNA are wrapped (Figure 4.3). The core histones are predominantly globular proteins, except for their protruding

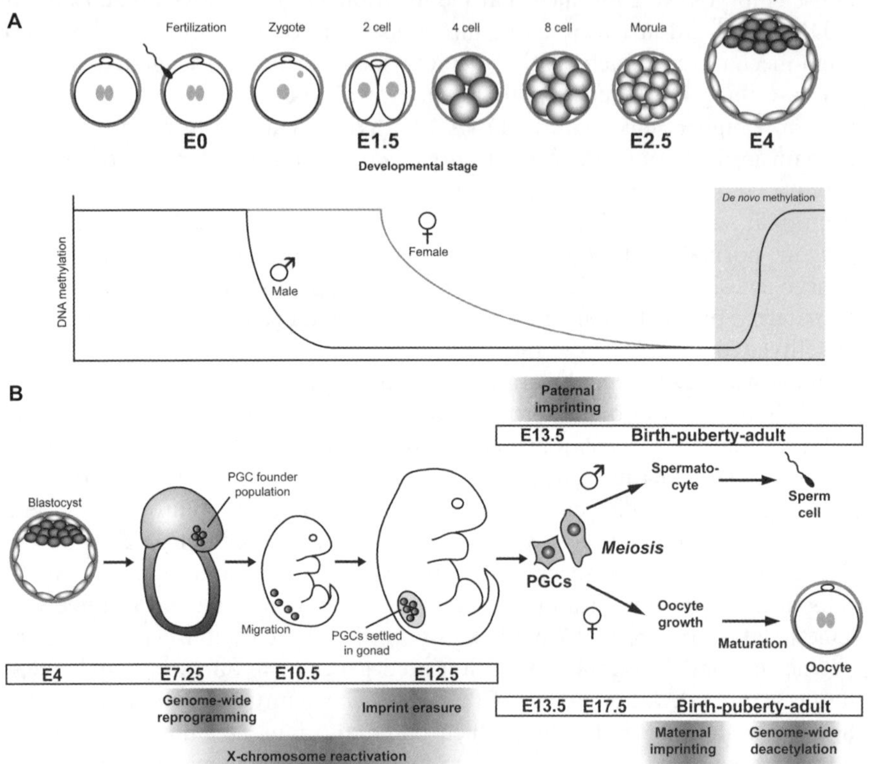

Figure 4.2   Genome-wide developmental resetting of epigenetic marks. (A) 4–8 hours post-fertilization, the sperm-derived pronucleus undergoes genome-wide active DNA demethylation. The oocyte, on the other hand, is protected from DNA demethylation at this stage. During the development of the fertilized zygote towards the blastocyst stage, the maternal genome undergoes passive DNA demethylation. Subsequently, *de novo* DNA methylation is established during the blastocyst development by DNMT3A and DNMT3B. Imprinted genes still retain their methylation marks during this pre-implantation development. (B) In primordial germ cells, a second phase of genome-wide demethylation occurs between day E7 and E13. Subsequently, DNA marks are re-established in a gender-specific manner. Parental imprinting occurs between E13 and E17, maternal imprinting perinatally. PGC, primordial germ cell.

N-termini that seem unstructured. These N-terminal tails are subject to a number of post-translational modifications. There are at least eight different modifications found in histones: acetylation, methylation, phosphorylation, ubiquitination, sumoylation, ADP ribosylation, deimination and proline isomerization. Modifications have been detected at more than 60 different amino acid residues. In particular, residues in the tails of H3 and H4, which are highly charged and tightly associated with DNA, are affected by chemical alterations.[21]

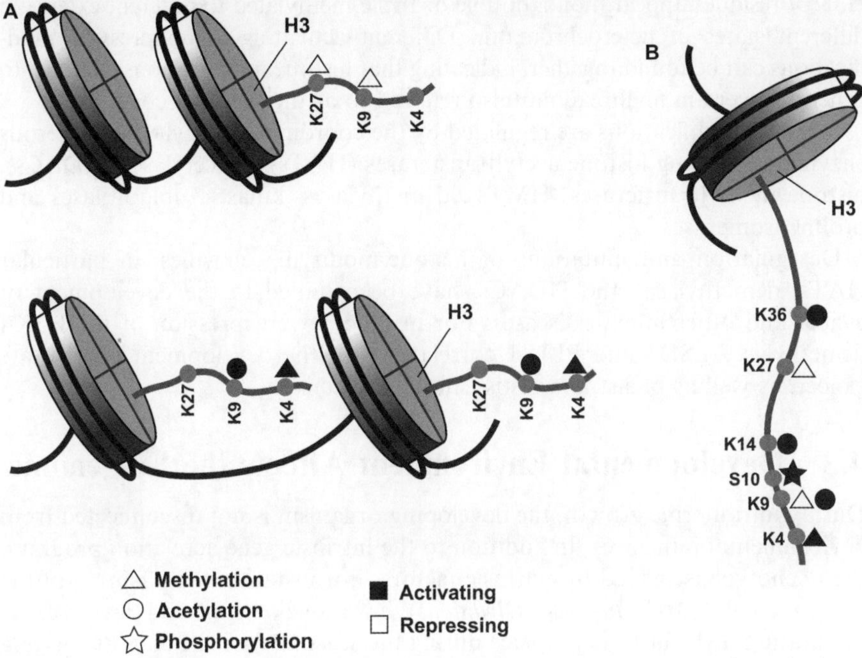

Figure 4.3   Histone modifications affect transcriptional activity. (A) Repressive and activating modifications of histone H3. Chemical modifications at H3 are involved in the organization of chromatin structure. Methylation of lysine at positions 9 and 27 (K9 and K27) in the unstructured N-terminus is associated with a closed and inactivated chromatin conformation, whereas acetylation of K9 and methylation of lysine 4 (K4) are activating and correlate with a loosened chromatin structure. (B) Landscape of activating and repressive H3 histone modifications. Illustrated are the best-characterized modifications.

Chemical modifications of histone proteins influence the higher-order chromatin structure by two different mechanisms. (1) They change the charge of the amino acid residues in a way that affects the contacts between histones of adjacent nucleosomes, and the interaction between histones and DNA. (2) They modulate the interaction of proteins, like chromatin remodeling enzymes, with chromatin.[22] Since the chromatin architecture and the grade of chromatin condensation determine the accessibility of the gene regulatory sequences to the transcription machinery, histone modifications have a strong influence on transcriptional activity. In this regard, histone acetylation is the most studied modification. Acetylation of specific lysine residues of H3 (H3K9 and H3K14) is associated with an open chromatin structure and transcriptional activity,[23,24] and the same appears to apply to certain serine phophorylations.[25] Methylation of lysine residues, on the other hand, cannot generally be associated with either repression or activation. Whereas methylation at lysine 4 in H3 (H3K4) is associated with transcriptional active chromatin, methylation at lysine 9 and 27 (H3K9, H3K27) associates with condensed and silent heterochromatin (Figure 4.3).[26] H3K9 and

H3K27 residues appear mono, double or triple methylated to different extents in different states of heterochromatin. Different activating and repressing modifications can be found together, indicating that histone modifications combine to generate a certain final readout (also referred to as the "histone code").[27,28]

Histone modifications are regulated by the coordinated activity of numerous enzymes, including histone acetyltransferases (HATs), deacetylases (HDACs), histone methyltransferases (HMTs), demethylases, kinases, ubiquitilases and proline isomerases.

Deregulation and mutations of histone-modifying enzymes, in particular HATs, demethylases and HDACs, have been linked to the development of cancer and other human diseases. For instance, overexpression of the H3K4 demethylases LSD1 and PLU-1 correlates with the development of prostate cancer, as well as breast and testis cancer, respectively.

4.3   Developmental Environment Affects the Epigenome

During intrauterine growth, the developing organism is not disconnected from environmental influences. In addition to the intrinsic gene activation program, the phenotype is shaped by extrinsic factors. For instance, the offspring of the meadow vole (*Microtus pennsylvanious*) have thicker fur and later sexual maturation if the birth takes place during the season when the days are getting shorter, compared to offspring of the same vole born in the spring when the days are getting longer.[29] Environment (day length) functions as a cue that guides the phenotype of the vole fetuses to a direction that assist survival after birth; the newborns are more likely to survive the winter if they have thicker fur and do not waste energy in sexual maturation and reproduction.

The capability to predictive adaptive responses, *i.e.* to develop different phenotypes from the same genotype in anticipation of life after birth, has been termed as phenotypic plasticity. Phenotypic plasticity gives the developing organism an opportunity to adjust to the challenges waiting in the outside world, regardless of the genotype, and these adjustments are believed to rely on epigenetic mechanisms. According to current thinking, epigenetic programming of the genetic material during early development in response to environmental cues can permanently alter the output of the genome, and this could subsequently have an effect on the risk of diseases. An interesting question is what kind of environmental cues can be interpreted by the developing fetus. It seems that at least the composition of the maternal diet, as well as presence of chemicals in the environment, can alter the epigenotype during development.

4.3.1   Early Life Environment, DNA Methylation and Susceptibility to Disease

4.3.1.1   Intrauterine Growth Conditions

In the 1980s, Barker *et al.* discovered a significant association between low birth weight and higher risk of cardiovascular disease in later life in humans.[30] Birth weight is a commonly used proxy for the growth conditions during intrauterine development: a low birth weight suggests nutritional deficiencies and/or placental defects. The association between adverse uterine growth conditions (*i.e.* low birth weight) and cardiovascular disease has been confirmed since Barker's original observation in various other studies encompassing both men and women, and different ethnicities.[31–33] These observations have collectively led to the concept of "developmental origin of health and disease" (DOHaD), also known as the Barker hypothesis. The hypothesis suggests that adverse conditions during intrauterine development, leading to lower birth weight, predispose for cardiovascular disease and metabolic syndrome later in life.

The Barker hypothesis has been tested in human populations that have been temporarily exposed to famine under exceptional circumstances. For example, during World War II, a part of the Dutch population suffered from severe famine termed the Dutch hunger winter, when the German occupation and an embargo on food transport combined with an unexceptionally harsh winter drove the people to starvation. The famine affected all people evenly, including pregnant women. Children conceived before or during the hunger winter displayed higher risk for hypertension, glucose and insulin intolerance, obesity and coronary heart disease in adult life.[34,35] Interestingly, the survivors of the Leningrad siege, who were likewise exposed to very severe famine, do not display increased risk of metabolic diseases as adults.[36,37] This discrepancy is largely attributed to the differing living conditions after the starvation period. While the Dutch population soon returned to prosperity, food shortage was a continued problem in the USSR even after the war. Hence, the environmental cues during early development matched better with the conditions after birth in Leningrad, while in the Netherlands this was not the case. It is the mismatch between the developmental cues and later living conditions that is believed to contribute to the development of metabolic diseases.

The very time-restricted nature of the Dutch hunger winter has enabled studies of the effect of the time point of starvation on later risk of disease. The incidence of obesity was only increased among adults who experienced the famine during early gestation.[35] Interestingly, major epigenetic programming takes place during early embryonic development (Figure 4.2A), and this time period might therefore be particularly sensitive for environmental influences. Indeed, among the Dutch hunger winter survivors, hypomethylation of the imprinted Igf2 gene and changes in the methylation of genes involved in metabolism is detected only in individuals exposed to the famine periconceptionally.[38,39] This observation provides the first evidence that the very early life conditions can provoke a persistent epigenetic change in humans.

Like in humans, maternal calorie restriction reduces birth weights in experimental animals.[40–42] In accordance with the mismatch hypothesis, and observations from the Dutch hunger winter survivors, growth-restricted pups gain more weight, have elevated plasma triglycerides and develop hyperglycemia and hyperinsulinemia when offered a high calorie diet directly after birth.[40] High calorie diet during gestation, on the other hand, leads to an increased birth weight and markers of cardiovascular disease in rodents as well.[41,43]

In addition, animal studies suggest that high birth weight increases susceptibility to develop mammary (breast) cancer in adult life.[44]

Interestingly, the birth weight is also reduced in pups whose mothers were protein restricted, but not calorie restricted.[40,45] Insufficient protein resources during early pregnancy could hamper placenta formation through effects on imprinted genes, such as the Igf2/H19 locus, which in mammals steer placental and fetal growth.[46] Genomic imprinting is achieved through DNA methylation and is therefore dependent on methyl donors, such as amino acids. Therefore, calorie restriction in general, and protein deficiency in particular, could lead to abnormal imprinting in early development and subsequent defects in placenta formation and fetal growth. In support of a connection between low protein intake and DNA methylation, developmental protein restriction has been shown to lead to hypomethylation and increased expression of genes coding for the glucocorticoid receptor and the peroxisome proliferator activated receptor alpha in the rat liver, as well as decreased expression of *DNMT1*.[43,44] In humans, a low ratio of protein to carbohydrates in the maternal diet reduces the weight of the placenta and the birth weight of the baby.[47,48] Hence, it seems that protein intake during pregnancy is important for the fetal growth rate (and later risk of disease) in humans too.

4.3.1.2 Stress

The nurturing that rat pups receive from their mother during the first week of life triggers an epigenetic change in stress-related genes in the brain.[49] If the pups are raised by a mother showing little affection (licking and grooming), the promoter of the glucocorticoid receptor in the hypothalamus and the glutamic acid decarboxylase gene promoter in the forebrain are hypermethylated.[49,50] If the pups born to an uncaring mother are transferred to a high-frequency grooming and licking mother, they will be programmed according to the foster mother's behavior. In the case of both genes, hypermethylation is associated with lower expression of the gene, which subsequently translates into a less effective negative feedback response through the glucocorticoid receptor in a stress situation, ultimately leading to an amplified stress response in the affected rat. Hence, the pups of an uncaring rat mother are behaviorally programmed to display higher stress behavior themselves, including lower maternal care, which will be transmitted on to their offspring.

In humans, some evidence about early life stress on cognitive function and epigenetic changes in the brain exists. For instance, the ribosomal RNA gene is hypermethylated and less expressed in the hippocampus of suicide victims that experienced abuse and neglect in their childhood compared to controls.[51] In schizophrenia patients, the level of SAM is increased and *DNMT1* is over-expressed in the prefrontal cortex, and the promoter of the reelin gene, which affects neuronal migration and cell signaling, is hypermethylated.[52,53] Furthermore, the survivors of extreme stress situations, such as the Dutch hunger winter and the great Chinese famine, display significantly higher frequencies of schizophrenia and other types of cognitive disorders as adults.[54–58] Collectively, these data suggest that early life environment, perhaps the combination of nutritional deficiencies and stress, can alter brain function and mental health in humans, perhaps through epigenetic mechanisms. However, it remains to be studied whether the methylation changes detected in *post mortem* brains are caused by early life environment or whether they arise due to the disease itself.

4.3.2 Endocrine Disruptors

4.3.2.1 The DES Story

Temporary exposure to single compounds during sensitive periods of development can have long-lasting effects on the health of the affected individual. Perhaps the most famous example of a single chemical with lifelong effects is diethylstilbestrol (DES). DES, a potent synthetic estrogen, was prescribed to pregnant women from the 1940s to the 1970s to prevent miscarriages and promote development of stronger babies. According to estimates, 8 million women used this drug before it was banned due to unusually high prevalence of vaginal clear cell carcinoma and reproductive track anomalies in the female offspring of the exposed mothers.[59,60] In rodent models, exposure to DES during fetal and neonatal development causes cancer, reproductive track abnormalities and altered gene expression profiles in different tissues.[61,62] Studies in mice have demonstrated that daily exposure to 2 μg of DES during the five first days of life causes hypomethylation of specific CpGs in the *lactoferrin* and *c-fos* genes, and exposure to DES *in utero* (10 μg kg^{-1} on days 9–16 of gestation) leads to hypermethylation of the *hoxa10* gene and increased expression of *DNMT1* and *DNM3B*, in the uterus of the pups.[63–65] These changes do not occur in mice exposed to DES as adults, demonstrating the high sensitivity of early developmental stages to chemical insults.[63,65] Although there are no data suggesting the presence of epigenetic changes in DES exposed humans, the data from experimental studies suggests that this compound can affect DNA methylation.

4.3.2.2 Other EDCs with an Epigenetic Connection

Similar to DES, several other compounds that target the endocrine system are suspected to disrupt the epigenetic code. The pesticide vinclozolin, that antagonizes androgen receptor activity, causes defects in spermatogenesis in rats if the exposure takes place during gonadal differentiation *in utero* (100–200 mg kg^{-1}, gestational days 8–15) and this phenotype is accompanied with a change in DNA methylation levels.[66] Exposure after gonadal differentiation (gestational days 15–20) does not affect spermatogenesis, suggesting that the time point of gonadal differentiation is the sensitive period to environmental influences. Interestingly, this is also the time period when major methylation changes occur in the maturing germ cells.

Methoxychlor, which is a pesticide used to replace the toxic DDT, targets imprinted genes in the male germ line in rats. When adult males are treated with

10 mg kg^{-1} methoxychlor on eight consecutive days, significant alterations in the imprinted *Gtl2*, *Peg1*, *Peg3* and *Snrpn* genes are induced. When the exposure is carried out during gonadal sex differentiation (days 10–18 of gestation), similar alterations in the imprinted genes are observed, and the changes are accompanied by decreased sperm count.[67] No changes in genomic imprinting were observed in somatic tissues, which suggests that methoxychlor specifically targets the male germ line.[67] Methoxychlor affects female rats too. When the compound is administered from gestational day 19 to postnatal day 7, at a dose of 100 mg kg^{-1}, hypermethylation of 10 different genes in the ovaries is detected, including the gene coding for estrogen receptor β.[68]

The environmental contaminant TCDD, which is a ligand for the aryl hydrocarbon receptor, interferes with the methylation of the imprinted *Igf2/H19* locus in mice. The *Igf2/H19* locus controls the growth of the placenta and the embryos, and typically *Igf2* is expressed from the paternal allele while *H19* is maternally expressed. If mouse embryos are exposed to TCDD from the one-cell stage to the blastocyst stage, and then implanted to unexposed recipient mice, the developing fetuses will display reduced *H19* expression along with increased methylation of the imprinting control region of the loci as well as increased methyltransferase activity. In addition, the TCDD fetuses are smaller than unexposed control fetuses on embryonic day 14.[69] As described above, imprinting is considered as a type of epigenetic mark that is not altered during early embryonic development but instead in the developing germ cells, where parent-of-origin specific allelic expression is set. It is therefore remarkable that TCDD disrupts the *H19/Igf2* imprinting control region methylation in early embryonic development. TCDD might also induce epigenetic changes later in fetal development. If rat fetuses are exposed to TCDD around the time period of initial mammary gland development, they display delayed mammary gland maturation after birth and develop more mammary tumors after dimethyl benzanthracene induced carcinogenesis as adults.[70] Although the authors of this study were able to show altered gene expression profiles in the mammary glands of the TCDD exposed rats on a protein level, they did not study the methylation of the control regions of these genes.

The agouti (A^{vy}) mouse is a model that exhibits distinct phenotypes in response to DNA hyper- and hypomethylation. The wild-type agouti (A) gene is normally expressed in a specific stage of hair growth in the skin of the mice, yielding a yellow stripe to the otherwise black hair of the mouse and leading to the typical pseudo-agouti (brown) coat color. The metastable A^{vy} allele of the agouti gene results from an insertion of a retrotransposon close to the transcription start site of the agouti gene. This retrotransposon contains a cryptic promoter that can promote ectopic expression of the agouti gene, *i.e.* expression outside its usual strictly controlled expression context. If the cryptic promoter of the metastable A^{vy} epiallele is hypomethylated, the agouti gene is expressed in a constitutive fashion, leading to yellow coat color, obesity and tumorigenesis. The development of the obese, yellow phenotype can be prevented by supplementing the diet of pregnant agouti mice with methyl donors, such as folic acid, vitamin B12, choline or betaine, which leads to methylation of the cryptic promoter.[71,72] Similar results are obtained if the diet is supplemented with genistein, a phytoestrogen present in soy.[73] However, another estrogen receptor ligand, the plasticizer bisphenol A, has the opposite effect in the agouti mice: when administered during early development, it caused hypomethylation, yellow coat color and obesity.[74]

Other EDCs with suspected effects on the epigenome include metals like arsenic, chromium and methylmercury, the organic solvent benzene, and persistent organic pollutants, as described below. Intake as well as serum and urine concentrations of arsenic correlate dose-dependently with global DNA hypermethylation, and with local hypermethylation of *p53* and *p16* genes, in blood cells in Indian and Bangladeshi adults.[75,76] Increased *p16* methylation is also seen in chromium-induced lung cancers in humans.[77] Methylmercury is a neurotoxicant that bioaccumulates in the food chains in wild life. The concentration of mercury in the brain of polar bears is dose-dependently and inversely correlated to DNA methylation on a global level, and this association is only seen in male polar bears.[78] There is a significant association between benzene in the breathing air and an aberrant methylation pattern in blood cells in Italian gas station attendants and traffic policemen. Specifically, benzene-exposed subjects displayed hypomethylation of repetitive elements and hypermethylation of the tumor suppressor *p15* in a dose-dependent manner.[79] Greenland Inuits are a population exposed to very high levels of persistent organic pollutants due to atmospheric transmission of these chemicals from industrialized countries. The concentration of DDT, DDE, oxychlordane, chlordane, mirex, polychlorinated biphenyl (PCB) congers and hexachlorobenzene in the blood of Inuits correlates with hypomethylation of repetitive elements in the blood cells.[80]

4.3.3 Transgenerational Effects

A largely discussed phenomenon in the field of epigenetics is the possible heritability of the chemically induced epigenetic changes from one generation to another. According to current knowledge, there are two developmental time points when the epigenetic marks (DNA methylation) are reset: during the germ cell maturation in the fetal gonad and immediately after fertilization in the zygote (Figure 4.2). It is unclear how chemically/environmentally induced changes in DNA methylation would persist through the reprogramming. Nevertheless, there are data suggesting that transgenerational inheritance of epigenetic phenotypes in mammals might indeed be possible. If the exposure is carried out during gonadal differentiation, both the mother (F0), the developing fetus (F1) and the primordial germ cells (F2) in the developing gonads of the fetus are directly exposed to the chemical (Figure 4.4). Therefore, only an epigenotype observed in the F3 generation without any further exposure would be evidence for inheritance. Some endocrine disruptors, and epigenotypes, seem to fulfill this criterion of inheritance.

Vinclozolin, an anti-androgenic pesticide, impairs spermatogenesis in male rats if the exposure occurs during gonadal sex differentiation, as described in

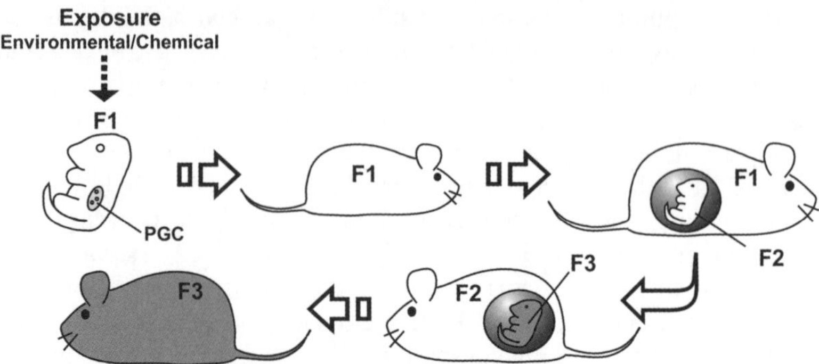

Figure 4.4 Inheritability of exposure. *In utero* exposure affects not only the pregnant mother (F0) and the embryo (F1) but also the primordial germ cells, and thus the second generation (F2). Therefore, only an epigenotype observed in the F3 and subsequent generations without any further exposure is evidence for inheritance.

the above section. The phenotype encompassing increased sperm apoptosis and decreased sperm motility can be observed in over 90% of male offspring all the way down to F4 generations, without any further exposure to the chemical.[66] The phenotype is associated with 52 significantly differently methylated regions in the sperm genome.[81] Several other chemicals, such as diethylstilbestrol, methoxychlor and bisphenol A, have been suggested to induce inherited epigenetic phenotypes, but clear evidence of inherited epigenotypes beyond the F2 generation have not yet materialized.[67,82]

Epigenetic changes stemming from the exposure of a fetus to a diet or a chemical through the mother can be readily reasoned because of the long, direct connection between the fetus and the mother. However, recent observations suggest that paternal environment can also lead to an epigenetic phenotype in the offspring. If rat fathers are kept on a high fat diet, they will gain weight and develop signs of metabolic dysfunction. When these males are mated to healthy females, the female offspring will have impaired glucose and insulin metabolism and changes in the gene expression profile and DNA methylation in the pancreas.[83] Alarmingly, some evidence of inheritability of metabolic phenotypes also exists in humans. For example, if the availability of food is high during the prepubertal growth period in men, their grandchildren have a higher risk of diabetes and shortened life span.[84,85] With the thought of worsening lifestyle habits and increasing obesity epidemics, significant research investments should be made to understand the parental influence on the metabolism and weight gain of the offspring.

4.4   Mechanisms by which EDCs Affect the Epigenome

While it is becoming clear that environmental factors, including chemicals, *can* induce epigenetic alterations, *how* they induce these alterations remains rather obscure. This is mainly due to the fact that some major questions regarding the regulation of the epigenome are still unsolved. For example, it is unclear how DNA methylation of specific CpGs is regulated, how active DNA demethylation is achieved or exactly how the different epigenetic mechanisms (DNA methylation, histone modifications, microRNAs) affect each other.

There are basically two possibilities how the epigenome can be affected: globally or locally. EDCs have been described to affect the epigenome on both levels. Mechanistically, global changes have been addressed, focusing on DNMTs and SAM in DNA methylation changes. On the other hand, mechanisms explaining gene specific changes in epigenetic marks have hardly been tackled yet.

4.4.1 Global Effects of EDCs on DNA Methylation

4.4.1.1 *EDCs Affect the Availability of SAM*

SAM (*S*-adenosylmethionine) is the methyl donor for virtually all methylation reactions in the cell, including methylation of DNA and histone tails. Processes that lead to depletion of SAM in the cell thus lead to a global reduction of methylation capacity. SAM is produced from methionine obtained through dietary sources, and this reaction is catalyzed by methionine adenosyltransferase (MAT). Methylation of a substrate results in conversion of SAM to *S*-adenosylhomocysteine (SAH), which is recycled back to SAM *via* homocysteine and methionine, thus constituting the methionine cycle (Figure 4.5).

EDCs have been shown to interfere with SAM metabolism on several levels. For example, arsenic, a toxic heavy metal, depletes the cellular SAM pool. In order to be excreted, arsenic is methylated in a series of oxidative methylation and reduction steps.[86] This biotransformation consumes both SAM and glutathione (GSH). Therefore, in the presence of excessive amounts of arsenic, the cell becomes depleted of SAM.[87] Furthermore, GSH is produced in the transsulfuration pathway that converts homocysteine to GSH *via* cysteine and γ-glutamylcysteine (Figure 4.5). Thus shunting of homocysteine into this pathway results in a reduction of SAH recycling to SAM, which in turn leads to genome-wide DNA hypomethyaltion.[88,89] Notably, GSH is not only important for excretion of arsenic but also for the metabolism of many other xenobiotics, including a number of EDCs such as dioxins and other persistent organic pollutants (POPs). Consequently, EDCs could affect global DNA methylation by inducing their metabolism and thus reducing SAM *via* GSH depletion.[90] In addition to consuming SAM in becoming methylated, arsenic also represses the formation of SAM. *In utero* exposure of mice to inorganic arsenic suppresses MAT and other enzymes involved in methionine metabolism on the mRNA level.[91] The aromatic hydrocarbon 3-methylcholanthrene (3-MC) is another EDC that reduces MAT mRNA and protein after *in vivo* exposure, and this effect was independent of the aryl hydrocarbon receptor (that mediates the xenobiotic response to 3-MC) and glucocorticoid signaling (which is an important regulator of MAT).[92]

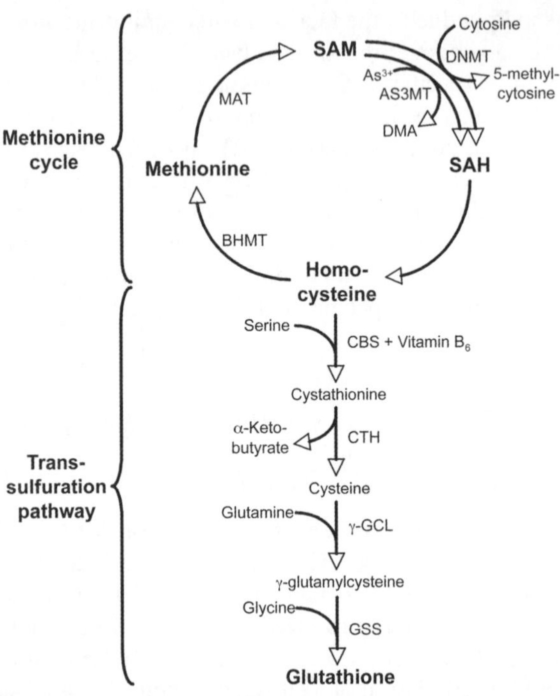

Figure 4.5   The methionine cycle and the transsulfuration pathway. AS3MT, arsenic methyltransferase; BHMT, betaine-homocysteine methyltransferase; CBS, cystathionine-β-synthase; CTH, cystathionase; DMA, dimethylarsinic acid; DNMT, DNA methyltransferase; gGCL, glutamate-cystine ligase; GSS, glutathione synthetase; MAT, methionine adenosyltransferase; SAH, *S*-adenosylhomocysteine; SAM, *S*-adenosylmethionine.

4.4.1.2   *EDCs Target DNMTs*

Exposure to EDCs can have long-lasting, even transgenerational, effects on the expression pattern of DNMTs. As described above, the fungicide vinclozolin reduces reproductive capacity in male rats if they are exposed during gonadal differentiation *in utero*.[66] This exposure leads to altered gene expression profile in the fetal testis, with 196 genes changed all the way down to the F3 generation, including reduced expression of *DNMT1*, *DNMT3A* and *DNMT3L* (a stimulator of DNMT3A and DNMT3B).[93] The pesticide methoxychlor (MTX), on the other hand, seems to have divergent effects on the different DNMTs: exposure between embryonic day 19 and postnatal day 7 resulted in increased *DNMT3B* levels, decreased (low dose MTX) or unchanged (high dose MTX) *DNMT3L* expression, and no changes for *DNMT3A* in the ovaries of female rats;[68] *DNMT1* was not measured in this study. DES has also been shown to induce *DNMT1* and *DNMT3B* levels in the uteri of *in utero* (embryonic day 9–16) exposed mice, whereas *DNMT3A* expression remained unaffected.[65] *In utero* exposure (day 12.5–19) to the plasticizer di-(2-ethylhexyl) phthalate (DEHP) resulted in increased expression of *DNMT1*, *DNMT3A* and *DNMT3B*.[94] Finally, arsenic modifies DNMT enzymatic activity in different cell lines, but the effect is not simply due to SAM depletion as cells incapable of methylating arsenic still showed reduced DNMT activity.[95–97] However, the results about the effect of arsenic on *DNMT* expression levels are somewhat conflicting: while one study reports increase of *DNMT1* mRNA, the follow-up study of the same group using different cells describes no changes.[95,96] On the other hand, two studies show a significant decrease of *DNMT1* and *DNMT3A* levels after arsenic exposure.[97,98]

In summary, these reports show that EDC exposure in early development can lead to changes in DNMT expression and/or activity in later life. However, some essential questions remain. For example, how are these changes linked to DNA methylation? Some of the studies show global hyper- or hypomethylation in their systems, but others show changes in DNA methylation on single promoters (Table 4.1). While it is plausible that global changes result from alterations in *DNMT* expression levels, it remains unclear how these alterations can affect specific gene promoters. Furthermore, it is unclear how EDCs induce changes in DNMT activity, and how long-lasting, even transgenerational, changes are sustained. *DNMT3B* expression has been shown to be positively regulated by estrogens in Ishikawa endometrial adenocarcinoma cells.[99] It is thus possible that *DNMT* mRNA levels are partly regulated by those hormonal systems that are affected by EDCs, and inappropriate stimulation or inhibition of these systems leads to changes in *DNMT* expression levels. However, this does not explain the sustained altered expression of *DNMTs* even in the absence of EDCs.

4.4.2 Local Effects of EDCs on the Epigenome through Nuclear Receptors

4.4.2.1 *Transient Chromatin Remodeling by Nuclear Receptors*

Many EDCs interfere with the signaling of ligand-induced transcription factors of the nuclear receptor family, such as the steroid hormone receptors estrogen receptor (ER) and glucocorticoid receptor (GR). Gene activation by these receptors results in complex chromatin remodeling processes, including changes in histone acetylation and methylation.[100] Recently, several studies have suggested that activation of gene transcription also coincides with changes in DNA methylation. The ERα target gene *pS2* exhibits periodic demethylation and remethylation, which coincides with cyclical ERα recruitment to the *pS2* promoter and expression of *pS2*.[101,102] Similarly, the vitamin D receptor, another member of the nuclear receptor family, seems to be involved in methylation of its target genes: treatment of cells with vitamin D induces methylation of the target gene *CYP27B1* accompanied with a repressed state of the gene, and treatment with parathyroid hormone leads to demethylation of the promoter and active state of the *CYP27B1* gene.[103] Finally, retinoic acid,

Table 4.1 Examples of EDCs that modify epigenome and phenotype.

Endocrine disruptor	Test system	Epigenetic effect	Phenotypic effect	Ref.
Diethylstilbestrol	Fetal exposure in humans Neonatal exposure in rats	Hypomethylation of *c-fos* and *lactoferrin* genes, hypermethylation of *hoxa10*, increased expression of DNMT1 and DNMT3B in the uterus (only observed in experimental animals)	Structural defects in the reproductive tract and higher incidence of cancer (observed both in experimental animals and humans)	59–62, 65
Vinclozolin	Fetal exposure during gonadal differentiation in rats	Fifty-two differently methylated regions in the sperm genome	Inheritable reduced reproductive capacity in males (increased sperm apoptosis and decreased motility)	66, 81
Methoxychlor	Fetal exposure during gonadal differentiation and exposure of adult male rats	Changes in the methylation of imprinted genes in sperm (not in somatic tissues) in adult and fetal exposure	Decreased sperm count (after fetal exposure only)	66–68
	Exposure during late gestation and early neonatal life	Hypermethylation of 10 genes and increased DNMT3B expression in the ovary in late gestation/ neonatal exposure	No information on ovarian phenotype	
TCDD	Exposure of early mouse embryos prior to implantation to recipient mothers	Increased methylation of the *Igf2/H19* control region, reduced *H19* expression, and increased methyltransferase activity in the fetuses	Decreased size of the fetuses	69, 70

	Exposure during initial mammary gland development in rats	No information on epigenetic lesions in the mammary gland	Altered mammary gland proteome, maturation and higher susceptibility to mammary carcinoma	
Bisphenol A	Dietary exposure of agouti mice starting periconceptionally and continued throughout gestation and lactation	Hypomethylation of metastable epialleles (A^{vy} and $Cabp^{IAP}$)	Shift of the coat color towards yellow agouti	74
Genistein		Hypermethylation of the metastable A^{vy} epiallele	Shift of the coat color towards brown pseudoagouti, higher likelihood of normal body weight	73
Arsenic	Humans exposed to high As drinking water	Dose-dependent hypermethylation of *p52*, *p16* and whole genome in blood cells	Exposure to arsenic is linked to skin lesions and cancer in humans	75, 76
Mercury	Polar bears	Dose-dependent whole genome hypomethylation in the brain of the males	Not known	78
Benzene	Gas station attendant and traffic policemen	Dose-dependent hypomethylation of repetitive elements and hypermethylation of *p15* in blood cells	Exposure to benzene is associated to acute myeloid leukemia in humans	79
Persistent organic pollutants	Greenland Inuits	Dose-dependent hypomethylation of repetitive elements in blood cells	Suspected connection to multiple adverse health effects in wild life, humans and laboratory animals	80

acting *via* the nuclear receptors retinoic acid receptor and retinoic X receptor, induces DNA demethylation of one single CpG site in its target gene *RET*.[104] However, these changes in DNA methylation, as well as the chromatin remodeling induced by nuclear receptors, are transient and thus do not explain the long-term effects of EDCs.

4.4.2.2   GR Induces Long-term DNA Demethylation

The GR promotes DNA demethylation of four specific CpGs of its target gene tyrosine aminotransferase (*Tat*).[105] *Tat* is a liver-specific gene and its transcription is turned on at birth, induced by sudden hypoglycemia after delivery. It can be activated by glucocorticoid activated GR, which binds to the glucocorticoid response element in the enhancer region of the *tat* gene.[106] In cultured human liver cells, short-term treatment with glucocorticoids leads to chromatin remodeling processes that are reversible upon hormone withdrawal.[107] However, prolonged glucocorticoid treatment (three days) results in DNA demethylation of all four CpGs in this region. This hypomethylation is stable for months after hormone withdrawal and results in faster and stronger transcriptional response to further glucocorticoid stimulation.[105] In experimental animals, this demethylation occurs during normal development before birth while the gene is still inactive. These DNA methylation changes might thus predispose the *tat* gene to neonatal induction by hypoglycemia, and factors interfering with these changes could affect the animal's response to hypoglycemia at birth.[105]

4.4.2.3   The ERs Promote DNA Methylation

Both ER isoforms have been suggested to promote gene repression by DNA methylation. ERβ binds to an estrogen response element in the paternally imprinted, maternally expressed *H19* gene in male rat germ cells. Furthermore, the receptor co-localizes and co-immunoprecipitates with DNMT1 in these cells.[108] Treatment with the ER antagonist tamoxifen results in reduced ERβ recruitment to the *H19* gene and in hypomethylation of the *H19* promoter.[108,109] The authors suggest that ERβ at this gene locus is responsible for the repression of the paternal allele of the *H19* gene and that chemicals inhibiting ER function, such as tamoxifen, can disturb this repression. However, in both studies, methylation data were not correlated with gene expression.

In a cell model, where breast progenitor cells were differentiated into breast epithelial cells, exposure of the progenitors to estradiol before differentiation leads to hypermethylation of a number of genes involved in tumor suppression in the epithelial cells.[110] This hypermethylation seems to be ERα mediated and concomitant with increased nuclear location of ERα and decreased gene expression in the epithelial cells.[110,111] Furthermore, exposure of the progenitor cells to different endocrine disruptors also changes gene expression of some of the genes that are hypermethylated after estradiol exposure.[111] The same authors report that exposure of the progenitor cells to DES leads to differential expression of some microRNAs. One of the promoters of the down-regulated microRNA showed silencing histone marks, recruitment of DNMT1 and DNA hypermethylation in the exposed cells.[112]

In summary, these studies suggest that liganded ERs are able to recruit DNMT1 to certain gene promoters and mediate DNA methylation and gene silencing. The finding that expression of some microRNAs is changed by EDCs could have further implications. For instance, it is possible that targeted microRNAs regulate members of the epigenetic machinery. Indeed, EDCs other than DES, such as some phthalates, polychlorinated biphenyls (PCBs) and bisphenol A, have been shown to down-regulate a number of microRNAs, including some that are predicted to target DNMTs.[112,113]

4.5   Conclusions and Future Perspectives

The interest in epigenetic phenomena has increased exponentially over the past few years. Currently, epigenetics is offered as an explanation for several common human diseases, such as cardiovascular disease and cancer, where genetic information seems to play only a minor role. In addition, accumulating evidence from studies in animals and humans suggest that even stress reactions, cognitive functions and mental diseases could involve an epigenetic component. Since extensive modifications of the epigenome occur in two phases during development, during fertilization and during maturation of primordial germ cells, these time points represent the sensitive time periods during which exposure to extreme conditions might have lifelong consequences. Indeed, more and more research focuses on epigenetics in embryonic, fetal and neonatal development to understand the basis of adult onset disease.

Endocrine disruptive chemicals are believed to be involved in the increasing rates of hormone-dependent diseases, such as fertility problems and breast cancer, in the industrialized parts of the world. These chemicals can bind to the steroid hormone receptors, triggering responses in the endocrine system that should only arise in response to the cognate ligands of the receptors, the hormones. Many hypotheses state that this inappropriate activation or repression of the hormonal system subsequently leads, for example, to the development of cancer. However, the recently observed connections between the nuclear hormone receptors, epigenetic changes and endocrine disruptors suggest the possibility of additional mechanisms underlying human diseases. Increasing amounts of scientific reports show that early developmental exposure to endocrine disruptors can induce epigenetic changes, leading to altered gene expression profiles in adult life. Such changes in gene expression, combined with continued exposure to environmental endocrine disruptors in adulthood, could play a significant role in the development of non-communicable diseases. However, as long as we do not understand the exact effects of EDCs on the epigenome, including the molecular mechanisms, the risk assessment of these chemicals remains a challenge. Significant efforts should thus be taken to unravel the correlation between exposure to EDCs, epigenetic changes and disease phenotypes, as well as to investigate at what exposure levels these changes take place.

Lastly, the evidence on hereditary epigenetic phenotypes is alarming. Animal studies have shown transmission of obesity phenotypes, both from the maternal and the paternal side, as well as reduced fertility phenotypes over several generations in response to transient exposure to endocrine disruptors. At the same time, reduced sperm counts, as well as obesity incidence reaching pandemic dimensions, are major problems in Western industrialized countries (and an increasing problem in developing countries). There are good reasons to believe that the environment where we live and the decisions we make not only affect our own health but the health of our children (and grandchildren) as well. Major research efforts should be taken to define the significance of epigenetic changes in fertility, cancer and metabolic disease in humans, to study what kind of environmental cues trigger the epigenetic alterations, and to find out how transgenerational inheritance could be prevented.

Acknowledgments

The authors would like to thank Dr Faiza Noreen for correcting the manuscript. The authors are supported by The Swedish Research Council (P.D.) and The Swiss National Research Foundation (J.R.).

References

1. L. N. Kolonel, D. Altshuler and B. E. Henderson, *Nat. Rev. Cancer*, 2004, **4**, 519–527.
2. *Epigenetic Mechanisms of Gene Regulation*, ed. V. E. A. Russo, R. A. Martienssen and A. D. Riggs, Cold Spring Harbor Monograph Series, Cold Spring Harbor Laboratory Press, Cold Spring Harbor, NY, 1996, vol. 32, pp. 2838–2844.
3. T. B. Johnson and R. D. Coghill, *J. Am. Chem. Soc.*, 1925, **47**, 2838–2844.
4. G. R. Wyatt, *Nature*, 1950, **166**, 237–238.
5. M. Ehrlich, M. A. Gama-Sosa, L. H. Huang, R. M. Midgett, K. C. Kuo, R. A. McCune and C. Gehrke, *Nucleic Acids Res.*, 1982, **10**, 2709–2721.
6. P. A. Jones and S. B. Baylin, *Nat. Rev. Genet.*, 2002, **3**, 415–428.
7. A. Eden, F. Gaudet, A. Waghmare and R. Jaenisch, *Science*, 2003, **300**, 455.
8. C. De Smet, A. Loriot and T. Boon, *Mol. Cell. Biol.*, 2004, **24**, 4781–4790.
9. Z. Siegfried and H. Cedar, *Curr. Biol.*, 1997, **7**, R305–R307.
10. R. Singal and G. D. Ginder, *Blood*, 1999, **93**, 4059–4070.
11. S. K. Ooi and T. H. Bestor, *Cell*, 2008, **133**, 1145–1148.
12. K. Rai, I. J. Huggins, S. R. James, A. R. Karpf, D. A. Jones and B. R. Cairns, *Cell*, 2008, **135**, 1201–1212.

13. M. Tahiliani, K. P. Koh, Y. Shen, W. A. Pastor, H. Bandukwala, Y. Brudno, S. Agarwal, L. M. Iyer, D. R. Liu, L. Aravind and A. Rao, *Science*, 2009, **324**, 930–935.
14. E. Li, T. H. Bestor and R. Jaenisch, *Cell*, 1992, **69**, 915–926.
15. A. Bird, *Genes Dev.*, 2002, **16**, 6–21.
16. W. Reik, W. Dean and J. Walter, *Science*, 2001, **293**, 1089–1093.
17. B. E. Bernstein, A. Meissner and E. S. Lander, *Cell*, 2007, **128**, 669–681.
18. S. Gopalakrishnan, B. O. Van Emburgh and K. D. Robertson, *Mutat. Res.*, 2008, **647**, 30–38.
19. A. P. Feinberg and B. Vogelstein, *Nature*, 1983, **301**, 89–92.
20. A. H. Ting, K. M. McGarvey and S. B. Baylin, *Genes Dev.*, 2006, **20**, 3215–3231.
21. A. C. D'Alessio and M. Szyf, *Biochem. Cell Biol.*, 2006, **84**, 463–476.
22. B. D. Strahl and C. D. Allis, *Nature*, 2000, **403**, 41–45.
23. K. Struhl, *Genes Dev.*, 1998, **12**, 599–606.
24. T. Y. Roh, S. Cuddapah and K. Zhao, *Genes Dev.*, 2005, **19**, 542–552.
25. N. Macdonald, J. P. Welburn, M. E. Noble, A. Nguyen, M. B. Yaffe, D. Clynes, J. G. Moggs, G. Orphanides, S. Thomson, J. W. Edmunds, A. L. Clayton, J. A. Endicott and L. C. Mahadevan, *Mol. Cell*, 2005, **20**, 199–211.
26. T. Kouzarides, *Cell*, 2007, **128**, 693–705.
27. T. Jenuwein and C. D. Allis, *Science*, 2001, **293**, 1074–1080.
28. S. L. Schreiber and B. E. Bernstein, *Cell*, 2002, **111**, 771–778.
29. T. M. Lee, L. Smale, I. Zucker and J. Dark, *J. Reprod. Fertil.*, 1987, **81**, 337–342.
30. D. J. Barker, C. Osmond, J. Golding, D. Kuh and M. E. Wadsworth, *Br. Med. J.*, 1989, **298**, 564–567.
31. C. Osmond, D. J. Barker, P. D. Winter, C. H. Fall and S. J. Simmonds, *Br. Med. J.*, 1993, **307**, 1519–1524.
32. B. Longo-Mbenza, R. Ngiyulu, M. Bayekula, E. K. Vita, F. B. Nkiabungu, K. V. Seghers, E. L. Luila, F. M. Mandundu and M. Manzanza, *J. Cardiovasc. Risk*, 1999, **6**, 311–314.
33. A. P. Candido, R. Benedetto, A. P. Castro, J. S. Carmo, R. L. Nicolato, R. M. Nascimento-Neto, R. N. Freitas, S. N. Freitas, W. T. Caiaffa and G. L. Machado-Coelho, *Eur. J. Pediatr.*, 2009, **168**, 1373–1382.
34. T. J. Roseboom, J. H. van der Meulen, C. Osmond, D. J. Barker, A. C. Ravelli, J. M. Schroeder-Tanka, G. A. van Montfrans, R. P. Michels and O. P. Bleker, *Heart*, 2000, **84**, 595–598.
35. T. Roseboom, S. de Rooij and R. Painter, *Early Hum. Dev.*, 2006, **82**, 485–491.
36. S. A. Stanner, K. Bulmer, C. Andres, O. E. Lantseva, V. Borodina, V. V. Poteen and J. S. Yudkin, *Br. Med. J.*, 1997, **315**, 1342–1348.
37. S. A. Stanner and J. S. Yudkin, *Twin Res.*, 2001, **4**, 287–292.
38. B. T. Heijmans, E. W. Tobi, A. D. Stein, H. Putter, G. J. Blauw, E. S. Susser, P. E. Slagboom and L. H. Lumey, *Proc. Natl. Acad. Sci. USA*, 2008, **105**, 17046–17049.

39. E. W. Tobi, L. H. Lumey, R. P. Talens, D. Kremer, H. Putter, A. D. Stein, P. E. Slagboom and B. T. Heijmans, *Hum. Mol. Genet.*, 2009, **18**, 4046–4053.

40. F. Bieswal, M. T. Ahn, B. Reusens, P. Holvoet, M. Raes, W. D. Rees and C. Remacle, *Obesity (Silver Spring)*, 2006, **14**, 1330–1343.

41. K. A. Lillycrop and G. C. Burdge, *Int. J. Obes. (London)*, 2011, **35**, 72–83.

42. Q. X. Yuan, J. Y. Zhou, L. P. Teng, C. P. Liu, J. Guo, L. J. Liu, W. De, K. F. Xu, X. D. Mao and C. Liu, *Horm. Metab. Res.*, 2010, **42**, 491–495.

43. J. A. Armitage, P. D. Taylor and L. Poston, *J. Physiol.*, 2005, **565**, 3–8.

44. G. C. Burdge, K. A. Lillycrop and A. A. Jackson, *Br. J. Nutr.*, 2009, **101**, 619–630.

45. K. L. Franko, A. J. Forhead and A. L. Fowden, *Nutr. Metab. Cardiovasc. Dis.*, 2009, **19**, 555–562.

46. A. L. Fowden, C. Sibley, W. Reik and M. Constancia, *Horm. Res.*, 2006, **65**(suppl. 3), 50–58.

47. D. M. Campbell, M. H. Hall, D. J. Barker, J. Cross, A. W. Shiell and K. M. Godfrey, *Br. J. Obstet. Gynaecol.*, 1996, **103**, 273–280.

48. K. Godfrey, S. Robinson, D. J. Barker, C. Osmond and V. Cox, *Br. Med. J.*, 1996, **312**, 410–414.

49. I. C. Weaver, N. Cervoni, F. A. Champagne, A. C. D'Alessio, S. Sharma, J. R. Seckl, S. Dymov, M. Szyf and M. J. Meaney, *Nat. Neurosci.*, 2004, **7**, 847–854.

50. T. Y. Zhang, I. C. Hellstrom, R. C. Bagot, X. Wen, J. Diorio and M. J. Meaney, *J. Neurosci.*, 2010, **30**, 13130–13137.

51. P. O. McGowan, A. Sasaki, T. C. Huang, A. Unterberger, M. Suderman, C. Ernst, M. J. Meaney, G. Turecki and M. Szyf, *PLoS One*, 2008, **3**, e2085.

52. H. M. Abdolmaleky, K. H. Cheng, A. Russo, C. L. Smith, S. V. Faraone, M. Wilcox, R. Shafa, S. J. Glatt, G. Nguyen, J. F. Ponte, S. Thiagalingam and M. T. Tsuang, *Am. J. Med. Genet. B*, 2005, **134B**, 60–66.

53. A. Guidotti, W. Ruzicka, D. R. Grayson, M. Veldic, G. Pinna, J. M. Davis and E. Costa, *Neuroreport*, 2007, **18**, 57–60.

54. E. S. Susser and S. P. Lin, *Arch. Gen. Psychiatry*, 1992, **49**, 983–988.

55. A. S. Brown, E. S. Susser, S. P. Lin, R. Neugebauer and J. M. Gorman, *Br. J. Psychiatry*, 1995, **166**, 601–606.

56. D. St Clair, M. Xu, P. Wang, Y. Yu, Y. Fang, F. Zhang, X. Zheng, N. Gu, G. Feng, P. Sham and L. He, *J. Am. Med. Assoc.*, 2005, **294**, 557–562.

57. E. J. Franzek, N. Sprangers, A. C. Janssens, C. M. Van Duijn and B. J. Van De Wetering, *Addiction*, 2008, **103**, 433–438.

58. M. Q. Xu, W. S. Sun, B. X. Liu, G. Y. Feng, L. Yu, L. Yang, G. He, P. Sham, E. Susser, D. St Clair and L. He, *Schizophr. Bull.*, 2009, **35**, 568–576.

59. P. Greenwald, J. J. Barlow, P. C. Nasca and W. S. Burnett, *New Engl. J. Med.*, 1971, **285**, 390–392.

60. A. L. Herbst, H. Ulfelder and D. C. Poskanzer, *New Engl. J. Med.*, 1971, **284**, 878–881.

61. R. R. Newbold, E. Padilla-Banks and W. N. Jefferson, *Endocrinology*, 2006, **147**, S11–S17.
62. S. Li, S. D. Hursting, B. J. Davis, J. A. McLachlan and J. C. Barrett, *Ann. N. Y. Acad. Sci.*, 2003, **983**, 161–169.
63. S. Li, K. A. Washburn, R. Moore, T. Uno, C. Teng, R. R. Newbold, J. A. McLachlan and M. Negishi, *Cancer Res.*, 1997, **57**, 4356–4359.
64. S. Li, R. Hansman, R. Newbold, B. Davis, J. A. McLachlan and J. C. Barrett, *Mol. Carcinog.*, 2003, **38**, 78–84.
65. J. G. Bromer, J. Wu, Y. Zhou and H. S. Taylor, *Endocrinology*, 2009, **150**, 3376–3382.
66. M. D. Anway, A. S. Cupp, M. Uzumcu and M. K. Skinner, *Science*, 2005, **308**, 1466–1469.
67. C. Stouder and A. Paoloni-Giacobino, *Reproduction*, 2011, **41**, 207–216.
68. A. M. Zama and M. Uzumcu, *Endocrinology*, 2009, **150**, 4681–4691.
69. Q. Wu, S. Ohsako, R. Ishimura, J. S. Suzuki and C. Tohyama, *Biol. Reprod.*, 2004, **70**, 1790–1797.
70. S. Jenkins, C. Rowell, J. Wang and C. A. Lamartiniere, *Reprod. Toxicol.*, 2007, **23**, 391–396.
71. G. L. Wolff, R. L. Kodell, S. R. Moore and C. A. Cooney, *FASEB J.*, 1998, **12**, 949–957.
72. R. A. Waterland and R. L. Jirtle, *Mol. Cell. Biol.*, 2003, **23**, 5293–5300.
73. D. C. Dolinoy, J. R. Weidman, R. A. Waterland and R. L. Jirtle, *Environ. Health Perspect.*, 2006, **114**, 567–572.
74. D. C. Dolinoy, D. Huang and R. L. Jirtle, *Proc. Natl. Acad. Sci. USA*, 2007, **104**, 13056–13061.
75. S. Chanda, U. B. Dasgupta, D. Guhamazumder, M. Gupta, U. Chaudhuri, S. Lahiri, S. Das, N. Ghosh and D. Chatterjee, *Toxicol. Sci.*, 2006, **89**, 431–437.
76. J. R. Pilsner, X. Liu, H. Ahsan, V. Ilievski, V. Slavkovich, D. Levy, P. Factor-Litvak, J. H. Graziano and M. V. Gamble, *Am. J. Clin. Nutr.*, 2007, **86**, 1179–1186.
77. K. Kondo, Y. Takahashi, Y. Hirose, T. Nagao, M. Tsuyuguchi, M. Hashimoto, A. Ochiai, Y. Monden and A. Tangoku, *Lung Cancer*, 2006, **53**, 295–302.
78. J. R. Pilsner, A. L. Lazarus, D. H. Nam, R. J. Letcher, C. Sonne, R. Dietz and N. Basu, *Mol. Ecol.*, 2010, **19**, 307–314.
79. V. Bollati, A. Baccarelli, L. Hou, M. Bonzini, S. Fustinoni, D. Cavallo, H. M. Byun, J. Jiang, B. Marinelli, A. C. Pesatori, P. A. Bertazzi and A. S. Yang, *Cancer Res.*, 2007, **67**, 876–880.
80. J. A. Rusiecki, A. Baccarelli, V. Bollati, L. Tarantini, L. E. Moore and E. C. Bonefeld-Jorgensen, *Environ. Health Perspect.*, 2008, **116**, 1547–1552.
81. C. Guerrero-Bosagna, M. Settles, B. Lucker and M. K. Skinner, *PLoS One*, 2010, **5**, e13100.
82. M. K. Skinner, M. Manikkam and C. Guerrero-Bosagna, *Reprod. Toxicol.*, 2011, **31**, 337–343.

83. S. F. Ng, R. C. Lin, D. R. Laybutt, R. Barres, J. A. Owens and M. J. Morris, *Nature*, 2010, **467**, 963–966.
84. L. O. Bygren, G. Kaati and S. Edvinsson, *Acta Biotheor.*, 2001, **49**, 53–59.
85. G. Kaati, L. O. Bygren and S. Edvinsson, *Eur. J. Hum. Genet.*, 2002, **10**, 682–688.
86. M. Vahter, *Toxicology*, 2002, **181/182**, 211–217.
87. J. F. Reichard and A. Puga, *Epigenomics*, 2010, **2**, 87–104.
88. K. Lertratanangkoon, C. J. Wu, N. Savaraj and M. L. Thomas, *Cancer Lett.*, 1997, **120**, 149–156.
89. J. F. Coppin, W. Qu and M. P. Waalkes, *J. Biol. Chem.*, 2008, **283**, 19342–19350.
90. D.-H. Lee, D. R. Jacobs, Jr. and M. Porta, *Environ. Health Perspect.*, 2009, **117**, 1799–1802.
91. J. Liu, Y. Xie, R. Cooper, D. M. Ducharme, R. Tennant, B. A. Diwan and M. P. Waalkes, *Toxicol. Appl. Pharmacol.*, 2007, **220**, 284–291.
92. M. V. Carretero, M. U. Latasa, E. R. Garcia-Trevijano, F. J. Corrales, C. Wagner, J. M. Mato and M. A. Avila, *Biochem. Pharmacol.*, 2001, **61**, 1119–1128.
93. M. D. Anway, S. S. Rekow and M. K. Skinner, *Genomics*, 2008, **91**, 30–40.
94. S. Wu, J. Zhu, Y. Li, T. Lin, L. Gan, X. Yuan, M. Xu and G. Wei, *Int. J. Toxicol.*, 2010, **29**, 193–200.
95. C. Q. Zhao, M. R. Young, B. A. Diwan, T. P. Coogan and M. P. Waalkes, *Proc. Natl. Acad. Sci. USA*, 1997, **94**, 10907–10912.
96. L. Benbrahim-Tallaa, R. A. Waterland, M. Styblo, W. E. Achanzar, M. M. Webber and M. P. Waalkes, *Toxicol. Appl. Pharmacol.*, 2005, **206**, 288–298.
97. X. Cui, T. Wakai, Y. Shirai, N. Yokoyama, K. Hatakeyama and S. Hirano, *Hum. Pathol.*, 2006, **37**, 298–311.
98. J. F. Reichard, M. Schnekenburger and A. Puga, *Biochem. Biophys. Res. Commun.*, 2007, **352**, 188–192.
99. M. Cui, Z. Wen, Z. Yang, J. Chen and F. Wang, *Mol. Biol. Rep.*, 2009, **36**, 2201–2207.
100. S. C. Biddie, *J. Neuroendocrinol*, 2011, **23**, 94–106.
101. S. Kangaspeska, B. Stride, R. Metivier, M. Polycarpou-Schwarz, D. Ibberson, R. P. Carmouche, V. Benes, F. Gannon and G. Reid, *Nature*, 2008, **452**, 112–115.
102. R. Metivier, R. Gallais, C. Tiffoche, C. Le Peron, R. Z. Jurkowska, R. P. Carmouche, D. Ibberson, P. Barath, F. Demay, G. Reid, V. Benes, A. Jeltsch, F. Gannon and G. Salbert, *Nature*, 2008, **452**, 45–50.
103. M. S. Kim, T. Kondo, I. Takada, M. Y. Youn, Y. Yamamoto, S. Takahashi, T. Matsumoto, S. Fujiyama, Y. Shirode, I. Yamaoka, H. Kitagawa, K. Takeyama, H. Shibuya, F. Ohtake and S. Kato, *Nature*, 2009, **461**, 1007–1012.
104. T. Angrisano, S. Sacchetti, F. Natale, A. Cerrato, R. Pero, S. Keller, S. Peluso, B. Perillo, V. E. Avvedimento, A. Fusco, C. B. Bruni, F. Lembo, M. Santoro and L. Chiariotti, *Nucleic Acids Res.*, 2011, **39**, 1993–2006.

105. H. Thomassin, M. Flavin, M. L. Espinas and T. Grange, *EMBO J.*, 2001, **20**, 1974–1983.
106. H. Sassi, R. Pictet and T. Grange, *Proc. Natl. Acad. Sci. USA*, 1998, **95**, 5621–5625.
107. T. Grange, L. Cappabianca, M. Flavin, H. Sassi and H. Thomassin, *Oncogene*, 2001, **20**, 3028–3038.
108. S. Pathak, R. D'souza, M. Ankolkar, R. Gaonkar and N. H. Balasinor, *Mol. Cell. Endocrinol.*, 2010, **314**, 110–117.
109. S. Pathak, N. Kedia-Mokashi, M. Saxena, R. D'souza, A. Maitra, P. Parte, M. Gill-Sharma and N. Balasinor, *Fertil. Steril.*, 2009, **91**, 2253–2263.
110. A. S. Cheng, A. C. Culhane, M. W. Chan, C. R. Venkataramu, M. Ehrich, A. Nasir, B. A. Rodriguez, J. Liu, P. S. Yan, J. Quackenbush, K. P. Nephew, T. J. Yeatman and T. H. Huang, *Cancer Res.*, 2008, **68**, 1786–1796.
111. P. Y. Hsu, H. K. Hsu, G. A. Singer, P. S. Yan, B. A. Rodriguez, J. C. Liu, Y. I. Weng, D. E. Deatherage, Z. Chen, J. S. Pereira, R. Lopez, J. Russo, Q. Wang, C. A. Lamartiniere, K. P. Nephew and T. H. Huang, *Genome Res.*, 2010, **20**, 733–744.
112. P. Y. Hsu, D. E. Deatherage, B. A. Rodriguez, S. Liyanarachchi, Y. I. Weng, T. Zuo, J. Liu, A. S. Cheng and T. H. Huang, *Cancer Res.*, 2009, **69**, 5936–5945.
113. M. Avissar-Whiting, K. R. Veiga, K. M. Uhl, M. A. Maccani, L. A. Gagne, E. L. Moen and C. J. Marsit, *Reprod. Toxicol.*, 2010, **29**, 401–406.

Phytoestrogens: Naturally Occurring, Hormonally Active Compounds in Our Diet

KRISTA A. POWER,*[a] OLIVER ZIERAU[b] AND
SHANNON O'DWYER[a,c]

[a] Guelph Food Research Centre, Agriculture and Agri-Food Canada,
93 Stone Road West, Guelph, Ontario, Canada, N1G-5C9; [b] Molecular Cell
Physiology and Endocrinology, Institute of Zoology, Technical University
Dresden, Germany; [c] Department of Human Health and Nutritional
Sciences, University of Guelph, Guelph, Ontario, Canada

5.1 Introduction

Phytoestrogens are estrogen-like di- or polyphenolic compounds produced in plants by way of the phenylpropanoid pathway.[1] While there are many types of phytoestrogens, this chapter will focus primarily on isoflavones, as well as highlight a more unique class of phytoestrogen, the prenylflavonoids. The interest in phytoestrogens as hormonally active dietary compounds stems from the fact that they can bind to estrogen receptors (ERs) α and β, activate ER signalling, and induce estrogenic responses in mammals, albeit with a much weaker potency compared to estrogen.[2] The physiological importance of this response will depend on several factors, including the timing of exposure to phytoestrogens (*e.g.* fetal or infant *vs.* adult; premenopausal *vs.* post-menopausal women).[3] This is because of the difference in the hormonal milieu

Issues in Toxicology No. 11
Hormone-Disruptive Chemical Contaminants in Food
Edited by Ingemar Pongratz and Linda Vikström Bergander

and sensitivity of target tissues to estrogenic stimuli at different developmental periods. With this in mind, this chapter focuses on the biological effects produced after exposure to phytoestrogens during more sensitive life stages: post-menopausal exposure to red clover isoflavones and infant exposure to isoflavone-rich soy formula. We have reviewed the *in vitro, in vivo,* and some human clinical studies highlighting the potential for these phytoestrogen exposures to induce estrogenic effects, which can result in both beneficial and adverse health effects. We have also reviewed the findings which demonstrate the estrogenic potential of the prenylflavonoids, a hops-derived class of phytoestrogens, which have only recently gained public attention as hormonally active dietary compounds.

5.2 Red Clover Isoflavones and Postmenopausal Health Effects

5.2.1 Isoflavone Composition, Metabolism, and Bioavailability

Red clover belongs to the plant family *Leguminosae,* along with many other leguminous plants, and is only one species of 250 in the *Trifolium* genus. It is an excellent source of isoflavones, in particular biochanin A (BioA) and for-mononetin (FOR); however, it also contains lower amounts of other iso-flavones, including genistein (GEN), daidzein (Daid), glycetin, irilone, prunetin, pseudobaptigenin, calycosin, and pratensein.[4,5] Until recently, the amounts and biological activities of BioA, FOR, GEN, and Daid have been the primary focus for research activities on red clover, while the other minor iso-flavones (irilone, prunetin, pseudobaptigenin, calycosin, and pratensein) have received little attention.[5] Red clover cultivation began centuries ago in Asia and Europe and gradually spread throughout the world, mainly due to its rapid growing capacity and ability to improve soil physiology. However, interest in red clover as a source of phytoestrogens commenced after the observation that sheep, grazing on red clover, had fertility problems, later referred to as "clover disease".[6,7] The plant characteristics and distribution of isoflavones varies dramatically in red clover due to growing conditions, soil quality, and genetic variability.[4,8,9] The isoflavones are found in all plant parts but in variable concentrations, with the highest proportion in the leaves, while the flower has a very low content.[4] Isoflavone-rich red clover extracts are now commercially produced and sold as "safe", natural, alternative therapies to hormone repla-cement therapy (HRT) for menopausal women, to help relieve menopausal symptoms and reduce the risk of associated diseases, including osteoporosis. It is thought that these natural estrogen-like alternatives would be void of the adverse estrogenic effects observed in some clinical trials using HRT, including increases in breast cancer and cardiovascular disease.[10] The results of such findings have led to a significant decline in the use of HRT and an increase in the use of herbal remedies by menopausal women.[11,12]

Once ingested, the red clover isoflavones, BioA and FOR, are demethylated in the gut to form GEN and Daid, respectively. These compounds can be further metabolized by the colonic microflora to produce more metabolites, such as 6-hydroxy-*O*-demethylangolensin and dihydrogenistein from GEN, and *O*-demethylangolensin (O-DMA) from Daid.[13,14] Furthermore, Daid can also be converted to its more estrogenic metabolite, equol, by the colonic microflora; however, this only occurs in about 1/3 of the population.[15] Limited studies, however, have been conducted to examine the bioavailability of the red clover isoflavones and metabolites after ingestion by humans. In one study, seven women (20–60 years) consumed a red clover supplement daily for five days, during which time urine was collected for isoflavone analysis. Based on the extract isoflavone profile, the subjects consumed 51.7 mg FOR, 84 mg BioA, 3.2 mg Daid, and 5.2 mg GEN. The authors also detected prunetin, pseudobaptigenin, and calycosin in the extract. While all compounds, as well as reduced, demethylated, and hydroxylated metabolites, were detected in the urine, Daid and GEN were the main urinary metabolites.[13] Furthermore, equol was produced in three of the subjects, which is in agreement with the current knowledge that only one-third of the population are equol producers.

In a more recent study, seven participants consumed 38.8 mg isoflavones (8.7 mg BioA, 18.9 mg FOR, 3.8 mg irilone, 1.3 mg Daid, 1.2 mg glycitein, and 0.2 mg GEN, as well as smaller amounts of prunetin and pseudobaptigenin) from a commercial red clover supplement and blood samples were collected 6.5 hours later.[5] Similar to the urinary patterns observed in the previously described study, plasma isoflavones indicate that although not fully metabolized, BioA and FOR are readily metabolized to GEN and Daid, respectively, with Daid levels (385 nM) being higher than GEN ($\sim$ 63 nM). Higher Daid levels may be due to the higher ratio of FOR:BioA in the extract used. In another study, using a red clover extract with higher BioA levels (24.5 mg) than FOR (16 mg) showed higher plasma GEN than Daid levels, collectively these findings highlight the importance of extract composition in the blood isoflavone profiles and potentially the biological effects induced by different red clover extracts.[16] Furthermore, irilone was also detected in the blood at levels similar to that of Daid ($\sim$ 351 nM), as well as prunetin and pseudobaptigenin, suggesting that these minor red clover isoflavones are absorbed and bioavailable.[5] Since research is now showing that irilone is practically resistant to microbial metabolism, further research is required to determine the biological role, if any, of these minor red clover isoflavones.[17]

5.2.2 Estrogenic Activities of Red Clover Isoflavones *In Vitro*

It is well known that the isoflavones GEN and Daid, as well as the enterometabolite equol, can bind and activate ERα and ERβ in numerous *in vitro* systems. It is also well known that these compounds have a higher preference for binding and activating ERβ compared to ERα.[18] With regards to the array of isoflavones and metabolites that can be produced after consumption of red clover, one study profiled their ER binding potency and categorized them in order of greatest binding affinity to ERα as: equol = GEN > 3-OH-GEN > Daid > dihydro-GEN > O-DMA = dihydro-Daid > FOR > BioA > 6-OH-O-DMA >

6-OH-Daid = angolensin; and to ERβ as: equol = GEN > 3-OH-GEN = Daid > dihydro-GEN > dihydro-Daid > BioA = 6-OH-Daid > O-DMA > FOR > angolensin = 6-OH-O-DMA.[19] In the same paper the potential to trans-activate ERα signalling was also determined and it was found that some of the compounds/metabolites were stronger activators of ERα signalling than they were ligand binders. The ranking for potency was equol = angolensin > GEN > BioA > O-DMA = FOR > Daid > dihydro-GEN > dihydro-Daid, while 3-OH-GEN, 6-OH-O-DMA, and 6-OH-Daid were unable to activate ERα signalling. From this study it is apparent that the parent compounds, FOR and BioA (the isoflavones present in the highest concentrations in red clover extracts), are weaker ER ligands, and for the most part, activators of ERα, than their metabolites. This is important with regards to the use of *in vitro* screening methods to determine the estrogenic potential of dietary/plant extracts, since the extract will be far more potent after being digested and metabolized. This also stresses the importance of *in vivo* testing, especially for compounds/extracts which become more active once metabolized. Furthermore, these results highlight that, depending on how red clover extracts are metabolized (*e.g.* equol producers *vs.* non-equol producers) or on the extract formulation (*e.g.* isoflavone composition: BioA-rich *vs.* FOR-rich extracts), variations in estrogenic response and biological effects may occur.

In addition to binding and activating the ER, studies have also determined if red clover extracts or the individual phytoestrogens can induce ER-mediated biological effects in various *in vitro* model systems. GEN and Daid, as well as equol, all have been shown to stimulate ER + MCF-7 breast cancer cell pro-liferation, reaching maximal stimulation at ∼ 1–10 μM.[20–23] In agreement with their potency ranking for activating the ER, while equol and Daid both can dose-dependently stimulate MCF-7 cell growth, equol is significantly more potent than Daid.[22] BioA (0.3–30 μM) has also been shown to stimulate the growth of MCF-7 cells; however, it has been reported that BioA is metabolized in certain cells to more estrogenic compounds (*i.e.* GEN), which may account for its estrogenic activity in MCF-7 cells.[24,25] Similarly, FOR can stimulate MCF-7 growth and ER-mediated gene expression; however, the stimulation only occurs at concentrations > 10 μM and continues to increase at con-centrations as high as 500 μM.[26] This appears to oppose the usual biphasic growth patterns induced by FOR metabolites, Daid and equol, which induces maximal cell proliferation at concentrations of 0.001–1 μM, after which cell proliferation starts to decrease.[23] Although not reported in the literature, it is possible that high FOR doses are partially metabolized by the cells to com-pounds (*e.g.* Daid or angolensin) that are more estrogenic and may stimulate cell growth. With regards to isoflavone extracts of red clover, 20 μg mL^{-1} of a red clover extract containing 30% isoflavones induced ERE reporter gene activity and ER-regulated gene expression (progesterone receptor) in MCF-7 cells, an effect that may have been due to GEN and BioA, since FOR and Daid, at the same dose (100 nM), were inactive in these assays.[27] The significance of screening extracts *in vitro*, such as in the MCF-7 cell proliferation assay, is questionable. As mentioned previously, the metabolites created after digestion and metabolism of red clover isoflavones (such as FOR and BioA) lead to the formation of an array of more estrogenic compounds; thus, the estrogenic potential of the extracts *in vitro* may be misleading.

5.2.3 Estrogenic Activities of Red Clover Extracts *In Vivo*

5.2.3.1 Animal Studies

The ultimate goal for a phytoestrogen-rich supplement to be used as an effective, safe, alternative to HRT is that it selectively targets specific estrogen-sensitive tissues, while having no effect on others. For example, an estrogenic response is desirable at tissues such as bone (to reduce loss of bone mass), vagina (to decrease vaginal atrophy and related symptoms), and for several menopausal symptoms (to reduce vasomotor symptoms), but would be undesirable at the breast and uterus, since some clinical studies have shown increased cancer rates in these tissues in women taking HRT.[28,29] The tissue-selective nature of red clover isoflavones is thought to be due to their higher binding affinity for ERβ compared to ERα. Since it is thought that ERα drives the major proliferative responses of estrogen in the uterus and breast, ERβ ligands are believed to be void of such effects. On the other hand, it is suggested that ERβ-containing tissues (*i.e.* bone, vagina) may see some benefits from red clover isoflavone supplements in postmenopausal women.

To determine the selective nature of red clover extracts, and thus the potential safe use of supplements in women, several animal models have been utilized to study "classical" estrogen sensitive tissues, including the uterus, vagina, and mammary gland. In one study, ovariectomized (OVX) Sprague-Dawley rats were given daily doses (37.5–112.5 mg kg^{-1} bw) of red clover extract isoflavones (57.3% FOR, 44% BioA, 2.3% Daid, and 5.6% GEN) by gavage, for 3 weeks.[30] The results showed that the red clover isoflavone extract dose-dependently increased uterine weight and vaginal cornification, two well-established biomarkers of estrogen action. In the mammary gland, however, the red clover extract did not exhibit any increase in ductal branching or alveolar budding, which were responses observed after estradiol treatment. Enhanced vaginal cornification has also been observed in another study in which rats were administered 100–200 mg kg^{-1} bw red clover isoflavones for 14 weeks, which may indicate that the red clover isoflavones can induce beneficial estrogenic effects on the vagina, resulting in a decrease in vaginal dryness and dyspareunia, common symptoms experienced by postmenopausal women.[31] On the other hand, a study using 1.5 mg kg^{-1} bw red clover isoflavones (isoflavone composition not reported), representing a physiologically relevant dose in women, for four weeks to OVX Wistar rats exhibited no growth or proliferative effects on the endometrium; however, an increase in ERα expression was evident, suggesting a weak estrogenic effect.[32] At a slightly higher concentration (2.4 mg kg^{-1} bw red clover isoflavones; 1:1 BioA to FOR), red clover isoflavones administered to OVX Spague-Dawley rats for 3 months had increased uterine weights compared to OVX controls.[33] This may suggest that estrogenic effects on the uterus are dependent on the dose of red clover extract consumed or the duration for which the supplement is taken.

Another major health concern for postmenopausal women is an increased risk of osteoporosis, which occurs with declining estrogen levels at menopause.[34,35] With the potential for red clover isoflavones to induce estrogenic effects on bone, several studies have tested an array of different red clover extracts for their effects on bone strength, bone mineral density (BMD), or bone turnover biomarkers in OVX rodents. In one study, OVX Sprague-Dawley rats were administered 100 and 200 mg kg^{-1} bw of red clover isoflavones (69% FOR, 24% BioA, 5.6% DAID, and 1.6% GEN) daily by gavage for 14 weeks.[31] The red clover treatments induced a higher femoral weight, density, and bone mineral content, compared to OVX controls. Red clover isoflavones also suppressed an increase in bone turnover biomarker, alkaline phosphatase (ALP), and increased the biomechanical strength of the tibia, indicating a reduction in fracture risk. In another study, OVX female Sprague-Dawley rats were fed diets supplemented with red clover extract (Menoflavon Forte; 200 mg tablet contains 80 mg isoflavones with ~ 1:1 of BioA and FOR) at a dose of ~ 2.4 mg kg^{-1} bw isoflavones, alone or in conjunction with a 16 mg alkaline treatment (calcium bicarbonate, sodium bicarbonate, magnesium and potassium carbonates, and disodium phosphate).[33,36] Red clover supplementation with or without alkaline treatment resulted in increased femoral biomechanical strength, increased bone mineral density, and an attenuation in the OVX-induced rise in bone turnover biomarker, osteocalcin, which was even further attenuated in the red clover + alkaline group. This suggests that while red clover treatment alone may reduce the risk of osteoporosis, combining it with an alkaline supplement may provide better protection. With regards to which red clover isoflavone(s) may be responsible for the beneficial effects on bone, a recent study compared the bone sparing effects of purified GEN, FOR, and Daid in OVX Spague-Dawley rats.[37] In this study, rats were administered 0.1, 1, 10 mg kg^{-1} bw FOR, or 10 mg kg^{-1} bw GEN or Daid, by intraperitoneal injections daily, for 9 weeks, and showed that all treatments increased bone mass in the tibia and spine, to a similar extent. This suggests that all red clover isoflavones, or their metabolites (*e.g.* equol), may reduce the risk of osteoporosis, and thus the isoflavone composition of the red clover extract may not be important for effects on bone health.

Collectively, these animal studies indicate that red clover isoflavones selectively target the reproductive tract and bone to induce estrogenic responses but may not stimulate mammary glands. While these studies may indicate the safe use of red clover for women with regards to breast cancer risk, no studies have been conducted to determine if the red clover isoflavones affect the growth of established breast tumors, as they do *in vitro* studies. The potential stimulatory effect on the uterus, however, highlights a safety concern for postmenopausal women taking red clover supplements (at least those with approximately an equal ratio of FOR and BioA), as estrogenic responses to the endometrium can led to hyperplasia, an increased risk factor for endometrial cancer.

5.2.3.2   Human Studies

Various studies have been conducted to determine if phytoestrogen-rich red clover extracts can induce beneficial effects in humans, particularly in women, as they are primarily marketed as natural safe alternatives to HRT. Various estrogen-related health outcomes have been investigated, including relief of menopausal symptoms, vaginal and uterine health, effects on mammary gland density, bone health, cognition and mood disorders, and risk factors for cardiovascular disease.[38–43] The results are highly variable, but may be explained by the differences in red clover supplement and isoflavone composition used, as well as the participant population and stage of menopause (perimenopausal *vs.* postmenopausal). The typical recommended daily dose of red clover isoflavones is 40–120 mg; however, extracts often have varying levels of the different isoflavones.[44–46] The following sections will describe some of the studies which have assessed the potential for different red clover supplements to affect menopausal symptoms, reproductive tract, and mammary glands, as well as risk factors for osteoporosis.

5.2.3.2.1 Menopausal Symptoms. With regards to relief of menopausal symptoms, several meta-analyses have been published to establish if red clover isoflavone extracts significantly reduce the number or frequency of hot flushes experienced by women, one of the major symptoms experienced in early menopause.[47–49] All meta-analyses included studies which were randomized, placebo-controlled trials, using Promensil or Rimostil as the source of red clover extract, resulting in the inclusion of only five or six trials. A Promensil tablet contains ∼40 mg isoflavones composed of ∼36.8% FOR, 59.7% BioA, 1.15% Daid, and 2.3% GEN, while a Rimostil tablet contains ∼30 mg isoflavones composed of 83% FOR, 8.3% BioA, and less than 3% of Daid and GEN, demonstrating the differences in total isoflavones/tablet as well as the vast differences in isoflavone composition. One meta-analysis included six red clover trials and found that only one study showed improvements in hot flushes compared to placebo controls.[49,50] After combining all studies, the mean difference in the number of daily hot flushes experienced by women taking red clover supplements compared with placebo was –0.44 (95% CI, –1.47 to 0.58), suggesting no protective effect. On the other hand, using the red clover supplement Menoflavon/MF11RCE, the effects on menopausal symptoms, vaginal health, and lipid metabolism were assessed in a double-blind, placebo-controlled, randomized cross-over design, trial. Fifty-three out of 60 women enrolled were assigned to ingest two tablets (40 mg isoflavones in each tablet consisting of 65% FOR, 26.3% BioA, 3.9% Daid, and 4.86% GEN) or a placebo control for 90 days.[51] Significant changes in vaginal cytology occurred in women after the red clover supplement compared to the placebo, including increased cornification, superficial cells, and karyopyknotic index. These changes are associated with beneficial estrogenic effects on vaginal health. There were also significant changes in menopausal symptoms, including hot flushes, night sweats, vaginal dryness, and dyspareunia, to name a few. This may indicate that the isoflavone composition of the red clover extract may be important in the effects induced on menopausal symptoms.

5.2.3.2.2 Effects on the Breast and Uterus. To assess the safe use of these extracts, a few studies have determined if uterotropic effects are induced after red clover supplement use. One study, using the MF11RCE supplement (80 mg isoflavones), investigated the effects of red clover isoflavone on endometrium thickness after 90 days in a double-blind, placebo-controlled, randomized cross-over design trial.[52] Unlike the effects observed in animal studies, there was a significant decrease in endometrium thickness, accompanied by an increase in testosterone levels, suggesting the safe use of this red clover supplement with regards to endometrial cancer risk.[30] This lack of uterotropic effect was also observed in a study which tested the effects of the red clover extract P-07 (Novagen, 40 mg isoflavones; 9.3 mg FOR, 25.7 mg BioA, 3.7 mg Diad, 4.3 mg GEN) in a placebo-controlled, double-blind, randomized trial in perimenopausal women.[53,54] In this study, endometrial thickness and cell proliferation from biopsy samples taken during the late proliferative phase of the menstrual cycle (days 7–11) were examined after 90 days of red clover supplementation. There were no differences observed between placebo and red clover groups, suggesting that the isoflavones, at the dose used, do not stimulate uterotropic effects. A dose–response study using the FOR-rich red clover extract Rimostil (28.5, 57, and 85.5 mg total isoflavones) for 6 months in postmenopausal women also showed no uterotropic effects, based on endometrial thickness, demonstrating that regardless of the isoflavone composition, no stimulatory effects on the uterus have yet to be identified after red clover supplementation.[55]

Two studies have addressed the role of Promensil in modulating breast cancer risk, as well as a recent meta-analysis combining these studies with other isoflavone sources (*i.e.* soy), all which demonstrate no significant effect on mammary gland density in postmenopausal women.[56–58] In agreement with results from previously mentioned animal studies, this suggests the potential safe use of red clover extracts for women, with regards to breast cancer risk.[68] However, in the human trials a dose of 40 mg isoflavones from Promensil were used, whereas doses as high as 120 mg red clover isoflavones, of varying isoflavone compositions, have been used in other studies not focusing on breast health.[30,53] Thus, to ensure no adverse estrogenic effects on the mammary gland, effects on breast tissue in women using higher doses of various red clover extracts should be assessed. Furthermore, there is no evidence to suggest the safe use of red clover isoflavones in women with existing breast cancer, and since isoflavones are known to increase the growth of ER + breast cancer cells, this should be assessed in future studies.

5.2.3.2.3 Effects on Bone Health. The potential for red clover extracts to reduce the risk of osteoporosis has also been observed in some human clinical trials. In one study, the FOR-rich red clover supplement Rimostil (28.5, 57, and 85.5 mg total isoflavones) was given to postmenopausal women for 6 months, which resulted in increased BMD in the forearm with the two highest doses.[55] Similar to the lack of effect of Rimostil at the lowest dose, another study using a 40 mg isoflavone dose of Rimostil for 50 days demonstrated no improvement in bone formation biomarkers, suggesting that the dose of isoflavones may be important for beneficial effects on bone, for this particular red clover supplement.[59] On the other hand, the Bio-A rich supplement Promensil has been shown to reduce bone mineral loss in the spine as well as increased bone formation biomarkers in postmenopausal women after 1 year of supplementation at a dose of 40 mg of isoflavones.[60] This may demonstrate the greater potential for BioA or GEN to reduce bone loss at lower doses compared to FOR or Daid, but also demonstrates that beneficial effects can be obtained by all red clover extracts, regardless of the isoflavone composition.

5.3   Soy Infant Formula and Potential Health Effects

Another population subgroup that can be exposed to high levels of phytoestrogens are soy formula fed infants. Soy is a rich source of isoflavones, but unlike red clover, soy primarily contains GEN, Daid, and glycetein, in glucoside form, namely genistin, daidzein, and glycetin, respectively. Soy formula prepared for infant consumption contains 32–47 µg mL^{-1} total isoflavones, with genistin accounting for $\sim$70% of the total isoflavones.[61] This results in an isoflavone intake ranging from 6 to 10 mg kg^{-1} bw, and plasma concentrations up to 5 µM, in infants up to 4 months of age, which is higher than levels of isoflavones in adults consuming a soy-based diet on a regular basis ($\sim$1 mg kg^{-1} bw).[61,62] It has also been shown that infants consuming soy milk/formulas excrete isoflavones in the urine, serum, and saliva, further demonstrating that the isoflavones are absorbed from the intestines and can be metabolized.[63,64] It is estimated that greater than 20 million infants have been exposed to soy formula in the USA since the 1970s.[65] The health effects associated with soy infant formula consumption is under current debate. Some studies suggest that beneficial effects are associated with early life exposure to soy formula, such as promoting normal growth and development, as well as reducing the risk of later breast cancer development.[65] However, others suggest that early exposure to estrogen-like compounds, such as isoflavones, can have detrimental effects on indices of reproduction, as well as increasing the risk of cancers of the reproductive tract.[66] Since the mass increase in soy formula usage only began around 30 years ago, only a few human studies have been conducted to determine the beneficial or detrimental long-term health effects of early isoflavone exposure; however, more elaborate studies have been conducted in animals.

5.3.1   Effects on Indices of Reproduction and Cancer Risk

Several studies utilizing newborn rodents have attempted to mimic formula exposure in human infants to determine the potential for soy isoflavones to induce estrogen-like effects (Table 5.1). Thus far, the majority of research in this field has been conducted by Newbold's group from the National Institute of Health.[67] In their earlier studies, they exposed newborn mice, from post-natal day (PND) 1 to 5, to GEN by subcutaneous injection. The dose used typically ranged from 5 to 50 mg kg^{-1} bw, which resulted in circulating GEN levels within range or a little higher than that seen in newborn infants consuming soy formula. They showed that GEN increased the size of the uterus and modulated ovary development in neonate females, as well as increased uterine cancer development, reduced fertility, and altered estrous cyclicity in the adult mice exposed to GEN during PND 1–5 only, effects consistent with estrogenization (Table 5.1). However, this model has attracted some criticism, primarily due to the fact that soy formula contains isoflavones as glucosides (genistin, daidzin, and glycitin) and not the more readily absorbed aglycone forms, GEN and Daid.[61] Furthermore, since formula is consumed orally, it has been criticized that subcutaneous injections will not accurately mimic infant feedings. Thus, newer studies have been conducted to demonstrate the effects of oral exposures to soy isoflavones, both in their glucoside and aglycone forms (Table 5.1).

Specifically, two recent papers highlight the effects of oral GEN or genistin on reproductive tract and reproductive capacity in mice after exposure for the first 5 days of life.[68,69] In one study, GEN was suspended in an oil emulsion solution and orally administered daily to pups at a concentration of 50 mg kg^{-1} bw.[69] This resulted in 3 µM serum concentrations of total GEN (conjugated + free) and ~1 µM of free GEN in pups 2–3 hours after the last GEN feeding. These concentrations fall within the range of that reported in infants fed soy formula.[61,70] Besides having elevated serum levels of GEN, several physiological outcomes were noted in these pups after 5 days of GEN feeding, as well as outcomes which were evident in adulthood (Table 5.1), indicating that the early exposure to soy infant formula levels of GEN induces changes which can have adverse events later in life. Similar effects were noted after neonatal oral exposure to genistin, indicating that the glucoside form is readily metabolized and absorbed to induce biological effects.[68] Although research is progressing in this field, criticisms still arise, primarily because soy infant formula contains other bioactive compounds, which may differently modulate the biological effects induced by purified GEN or genistin alone.[65] However, in a recent NIEHS Sister Study using self-administered family history questionnaires, the incidence of uterine leiomyometra (fibroid-benign smooth muscle tumors) in adult women was correlated with various *in utero* and early life events.[71] The study found that women who consumed soy infant formula in the first two months of life had a higher risk of uterine fibroids diagnosis in early adulthood (RR = 1.25; 95% CI, 0.90–1.73). Furthermore, another study showed that girls who consumed soy formula had a higher maturation index of vaginal cells compared to those fed breast or cow milk.[72] Collectively, both human and animal studies, thus far, support the hypothesis that early life exposures to dietary estrogens can induce effects on the reproductive tract; however, future long-term studies are needed to fully understand the impact of early soy isoflavone exposure.

Table 5.1 Summary of the effects of early life exposure to isoflavones in animal models.[a]

Treatment and dose	Model	Isoflavone effects in the neonate	Isoflavone effects in the prepubertal → adult rodent	Ref.
GEN (50 mg kg^{-1} bw subcutaneously) (subQ)	CD-1 female mice treated PND 1–5	Increase uterus weight	No corpora lutea in ovaries of GEN mice while 100% corpora lutea in controls; no ovarian tumors induced by GEN; uterine tumors developed in 35% GEN mice while none in controls	112
GEN (0.5, 5, or 50 mg kg^{-1} bw subQ)	CD-1 and C57Bl/6 female mice treated PND 1–5	0.5 and 5 GEN doses increased ovarian granulosa cell ERα and ERβ expression at PND 5, while 50 GEN dose decreased ERα and ERβ in CD-1 mice	GEN increased MOF in CD-1 and C57Bl/6 dose-dependently at PND 19 mediated by ERβ; 0.5 GEN dose increased number ovulated oocytes	113
GEN (0.5, 5, or 50 mg kg^{-1} bw subQ)	CD-1 female mice treated PND 1–5	–	ND in timing of VO; ND in serum progesterone or estradiol at PND 19; altered estrous cycling at 2 and 6 months of age, demonstrating either persistent diestrous or estrous. Altered fertility: 50 GEN dose did not deliver any live pups due in part to reduced numbers of successful pregnancies, numbers of uterine embryo implantation sites, reduced ovarian corpora lutea; 5 GEN dose had reduced number of live pups when mated at 2, 4, and 6 months of age	114

GEN (50 mg kg^{-1} bw subQ)	CD-1 female mice treated PND 1–5	Increased unassembled ovarian follicles and reduced primordial and primary follicles at PND 4, suggesting MOF may be a result of a disruption in ovarian differentiation	–	115
GEN (12.5, 20, or 25 mg kg^{-1} bw subQ or 25, 37.5, or 75 mg kg^{-1} bw orally)	CD-1 female mice PND 1–5	SubQ GEN: 20 and 25 doses increased uterus weight at PND 5; 25 dose increased uterine LF expression at PND 5 Oral GEN: 75 dose increased uterus weight at PND 5	Oral GIN: Dose-dependent increase in ovarian MOF at PND 19; delayed timing of VO with 50% in 37.5 dose opening at PND 31 *vs.* PND 29 for controls; abnormal estrous cycling (primarily prolonged estrous phase) at 2 months dose dependently. Reduced fertility after breeding 2, 4, and 6 month old mice: number of live births and number of pups born reduced in 25 and 37.5 dose groups at 6 months; dose-dependent delay in parturition at 2, 4, and 6 months	68
GIN (6.25, 12.5, 25, 37.5 mg kg^{-1} bw orally (converted to GEN dose equivalent)		Oral GIN: 25 and 37.5 doses increased uterus weight at PND 5; 37.5 dose increased uterine LF expression at PND 5; serum GEN at PND 5 = 5 µM		
GEN (50 mg kg^{-1} bw orally)	C57Bl/6 mice PND 1–5	ND bw; decrease in thymus weight at PND 5; increased uterus weight at PND 5 and reduced uterine luminal epithelial progesterone receptor expression; increased number of mice with MOF at PND 5; serum GEN at PND 5: ~1 µM	ND bw; ND timing of VO; GEN increased number of mice with ovarian MOF at 4 months; GEN decreased number of complete estrous cycles at 6 months due to prolonged diestous phase; no effects on fertility at 6 months of age (litter size and number/sex of pups)	69

Table 5.1 (*Continued*)

Treatment and dose	Model	Isoflavone effects in the neonate	Isoflavone effects in the prepubertal → adult rodent	Ref.
GEN (5 mg kg^{-1} bw), Daid (2 mg kg^{-1} bw), GEN + Daid subQ	CD-1 mice PND 1–5	Serum isoflavones at PND 5 in GEN (2.61 µM), Daid (1.07 µM), GEN + Daid (2.86 and 1.18 µM) treated mice, respectively. No detectable equol levels	At 4 months of age: ND in body weights. Males: ND spine or femur BMD; Daid increased strength (peak and yield load) at the femur neck. Females: all treatments increased spine BMD while only Daid increased femur BMD; Daid increased strength (femur and spine peak load), while all treatments increased femur strength (stiffness)	73
GEN + Daid (5 + 2 mg kg^{-1} bw, respectively, subQ)	CD-1 mice PND 1–5 and gonadomectomized at 4 months of age	–	32 weeks after gonadidectomy, GEN + Daid treatment from PND 1–5 increased female bw with ND in males; GEN + Daid increased spine and femur BMD and strength (peak load) in females only	76

[a]BMD = bone mineral density; bw = body weight; Daid = daidzein; GEN = genistein; GIN = genistin; LF = lactoferrin (estrogen-regulated protein); MOF = multioocyte follicles; ND = no difference; PND = postnatal day; VO = vaginal opening.

5.3.2 Effects on Bone Health

To support the hypothesis that there are beneficial effects of soy infant formula on growth and development, Ward *et al.* demonstrated that treating CD-1 mice from PND 1–5 with GEN, Daid, or a combination of GEN and Daid (at a ratio similar to that seen in soy infant formulas), improved BMD and strength at 4 months of age, with Daid inducing the greatest effects (Table 5.1).[73] This study showed that combining GEN and Daid was less effective than Daid alone, demonstrating that GEN may be interfering with the beneficial effects of Daid. Furthermore, the effects in female mice were more prominent than those observed in male mice. These studies indicate that different isoflavones are more effective than others with regards to their ability to modulate bone microstructure and fracture risk. Interestingly, while GEN and Daid serum levels were equivalent to those seen in newborns fed soy infant formula, serum levels of equol were undetectable in mice after Daid treatment. From a previous study, the ability of adult CD-1 mice to convert dietary Daid to equol was confirmed, suggesting that unlike adult mice, newborn CD-1 mice may not possess the appropriate bacteria to convert Daid to equol.[74] This is also supported in human studies, in which equol is rarely detected in biological samples taken from soy formula fed infants.[63] Thus, since it has been suggested that the beneficial effects of Daid are primarily due to equol, these findings suggest that Daid, or other metabolites of Daid, are directly responsible for the beneficial effects on bone when treated early in life.[75] In a later study, the group also determined if the combined GEN + Daid treatment administered at PND 1–5 could prevent bone loss and fracture risk in OVX or orchidectomized mice at 4 months of age (Table 5.1).[76] At 32 weeks after ovariectomy, the GEN + Daid treatment enhanced female bone mineral and strength at the spine and femur, but not in males. Micro-computed tomography of the vertebrae and femur neck indicated that this effect was due to enhanced cortical thickness and improved trabecular network compared to non-treated controls. Taken together, these animals studies suggest that early soy isoflavone exposure may promote bone health later in life; however, the effects may be gender specific, as well as isoflavone-type specific.

5.4 Prenylflavonoids

5.4.1 Chemical Structures and Sources of Prenylflavonoids

Prenylflavonoids or prenylated flavonoids are a sub-class of flavonoids, but in contrast to the intensively investigated isoflavone-type phytoestrogens (*e.g.* GEN and Daid), there is still relatively little known about the effects and actions of prenylated flavonoids. By far the most intensively studied prenylflavonoid is 8-prenylnaringenin (8PN). Beside 8PN, xanthohumol (XN), isoxanthohumol (IX), 6-prenylnaringenin (6PN), and 6-[1,1-dimethylallyl)]naringenin (6DMAN) have been studied to a certain degree (Figure 5.1). Most of these prenylated flavonoids, and a number of others, are found in hops, *Humulus lupulus* (L.). The major food sources of prenylflavonoids are hop-processed products such as

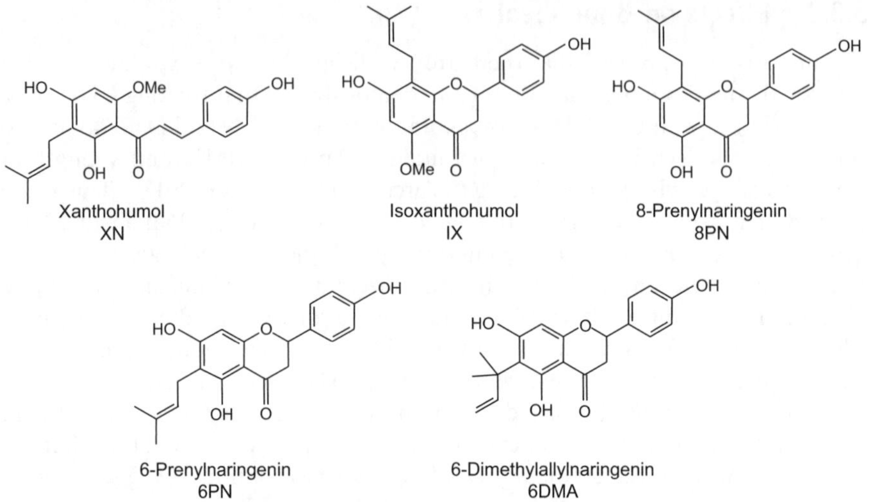

Figure 5.1 Chemical structures of several prenylated flavonoids.

beers and ales.[77–79] Furthermore, in the last few years a number of dietary supplements have been placed on the market which contain hops or isolated hop ingredients, and additional preparations with hop ingredients are planned to be launched by a number of producers.[80–82] Nevertheless, the dietary relevance of prenylflavonoids is mainly linked with the ingestion of beer.[83]

The detection of prenylated flavonoids in hops was triggered by anecdotal data on the interference of hop products with female reproduction. It was known for years that female hop pickers often had disruption of their menstrual cycle during the hops harvest.[84] This knowledge was confirmed in 1999 when Milligan *et al.* discovered a number of prenylflavonoids in hops and beer and highlighted 8PN as the most potent phytoestrogen in hops.[85] In Czech beer, the concentration of IX varied from 0.17 to 6.14 µmol L^{-1} beer sample, while the XN was between 10- to 100-fold lower.[86] Depending on the type of beers tested, 8PN concentration can vary, ranging from not detectable to 17.9 µg/L beer.[87] Besides hop products, a number of prenylated flavonoids have also been isolated from Japanese mulberry tree, *Morus alba* L. The fruits of mulberry trees are eaten, often dried or made into wine, but there are no reports if these sources of prenylated flavonoids exhibit any estrogenic activity.[88] Before 8PN was described as a hop ingredient, it was first isolated from another source and named 8-isopentenylnaringenin, which was found in a Thai crude drug, together with other prenylflavonoids. This traditional drug derives from the duramen of *Anaxagorea luzonensis* A. Gray, and the authors demonstrate estrogenic effects of 8PN.[89,90] Furthermore, prenylated flavonoids could be isolated out of a number of other plants with mainly no relevance for human nutrition, for example *Marshallia grandiflora* (Beadle & F. E. Boynt) out of the family of *Asteraceae* or *Sophora tomentosa* (L.) out of the family of *Fabacea*.[91,92]

5.4.2 Estrogenic Effects of Prenylflavonoids *In Vitro*

For several prenylflavonoids, such as 8PN and 6DMAN, relative low toxicity and weak cytostatic properties have been reported.[93] These findings made them even more attractive for drug development. On the other hand, flow cytometric and biochemical analysis showed that 8PN could result in cell cycle arrest and was able to induce apoptosis in MCF-7 human breast cancer cells.[83] Hop extracts and individual hop constituents, including 8PN and 6PN, have also been positively tested for their estrogenicity in different ERα and ERβ binding assays.[94] In contrast to these mammalian *in vitro* assays, 8PN and 6DMAN showed no estrogenicity in fish *in vivo* test systems.[95] Interestingly, in a neuronal cell model, 8PN showed strong concentration-dependent estrogenic activity, while 6DMAN neither showed estrogenic nor anti-estrogenic activity in the ERβ expressing neuronal cell model.[96] While IX is *in vitro* moderately estrogenic, no estrogenic effects could be detected for XN.[79,97] Furthermore, data from experiments with Caco-2 cells propose that specific binding of XN to cytosolic proteins of the intestinal epithelial cells results in relatively low bioavailability of an orally administered substance.[98] From experiments with human liver microsomes it was suggested that the estrogenicity of hop constituents *in vivo* depends in part on metabolic conversion, which may vary on an individual level.[99] Other experiments on *in vitro* metabolism of 8PN by human liver microsomes resulted in up to 12 metabolites which could be identified. This biotransformation affected the prenyl group as well as the flavanone skeleton.[100] 8PN can be detected as a product of IX demethylation, while an analogous demethylation reaction for XN was not observed.[97] However, Pang *et al.* concluded that XN is a precursor for IX and 8PN.[98] The 8PN concentration in beer was believed to have a relatively low impact on human health, until it was demonstrated that the activity of human gut flora could increase the exposure concentration up to a factor of 10. Therefore IX, the major constituent of the prenylflavonoids in many beers, might result, depending on the beer consumption, in an increased *in vivo* level of 8PN and thereby very well have an influence on human health.[101] IX can also be metabolized in the human liver to form 8PN.[99] Further, it could be demonstrated in independent *in vitro* test systems that the two major human metabolites of 8PN are still of relevant estrogenic potency.[102] These results also hint to a longer activity of 8PN.

Interestingly, the prenylation has a compound-specific effect on the estrogenicity, which differs enormously between naringenin and genistein. While 8-prenylnaringenin is the most estrogenic phytoestrogen, its counterpart 8-prenylgenistein is much weaker than 8PN and, more interestingly, even much less active than the unsubstituted genistein.[103] Besides that, prenylated flavonoids can act not only *via* the ER but also through the androgen receptor, as distinct *in vitro* anti-androgenic activity could be demonstrated for 8PN and 6DMAN.[104]

5.4.3 Estrogenic Effects of Prenylflavonoids *In Vivo*

In vivo experiments suggest that 8PN has to be regarded as a pure estrogen agonist.[105] In a short-term experiment on time dependency of uterine responses, the uterine wet weight and estrogenic gene expression induced by 8PN mimics that of E2-induced responses in the OVX rat.[106] 8PN and its metabolites have been detected in rat plasma, liver, and mammary gland, which demonstrates bioavailability.[107] In *in vivo* studies, 8PN has an anti-atherosclerotic profile which may demonstrate its potential for the prevention of estrogen deficiency associated cardiovascular diseases.[108] After long-term treatment of three months with 8PN, decreased bw and an increased uterus weight were observed, as well as with estradiol benzoate.[108] In another three month experiment, OVX rats treated with 8PN share many effects in the uterus, vagina, and mammary gland with estradiol; at the same time, differences in the mechanism of action seem to exist.[109] Other experiments in OVX rats, as a model for hormone-dependent osteoporosis, demonstrated that 8PN can completely protect from ovariectomy-induced bone loss.[110] In addition to the effects induced in females, 8PN can affect the male organism. As shown in a transgenic reporter mouse model (ERE-Luc mice), 8PN can exhibit pronounced estrogenic activity in the prostate.[110]

A very limited number of clinical studies have been performed for prenylated flavonoids so far. The potential health effects of prenylflavonoids on breast tissue in comparison to E2 were assessed in a dietary intervention study. The result of this study indicated that low doses of prenylated flavonoids are not likely to produce estrogenic effects in breast tissue.[111]

5.5 Summary

The potential for phytoestrogens to act as hormonally active compounds in our diet has been reviewed in this chapter, with particular reference to the isoflavones and prenylflavonoids. It is known that these phytoestrogens can bind to ERα and ERβ, and induce estrogenic effects in various *in vitro* and *in vivo* model systems. It is also known that the estrogenic potency of the phytoestrogens will vary, depending on the chemical structure and metabolites formed. Therefore, depending on the composition of the phytoestrogens consumed and the nature of colonic microfloral population, differences in the profiles of bioavailable phytoestrogens and biological effects can be obtained. This chapter also placed emphasis on describing the current knowledge on phytoestrogen action in populations that may be more susceptible to estrogenic stimuli. Postmenopausal women have low circulating estradiol levels, and thus increased exposure to plant estrogens may induce significant biological effects. Isoflavone-rich extracts derived from red clover plants are used by women to reduce menopausal symptoms and associated diseases, such as osteoporosis. Although many *in vitro* and *in vivo* studies demonstrate the estrogenic potential of red clover isoflavones, thus far there is a lack of high-quality randomized placebo-controlled clinical trials to make definitive conclusions as to their beneficial health effects in women, as well as their potential to induce adverse effects in tissues such as the breast and uterus. Furthermore, the large variability between supplements, with regards to dose and isoflavone composition, as well as the unknown biological effects of the minor isoflavones found in red clover supplements, stresses the need for proper chemical and biological characterization of these products. Future studies should be conducted to determine the extent to which isoflavone dose and composition matter, with regards to specific health outcomes in menopausal women. This chapter also highlighted the *in vivo* evidence demonstrating the estrogenic potential of soy infant formulas. From these studies it is clear that newborn rodents exposed to GEN, or its glycoside form genistin, can induce adverse effects on indices of reproduction and on the reproductive tract. Importantly, the circulating GEN levels observed in rodent studies are similar to those found in infants fed soy formula. Whether these findings in animals translates to humans is yet unknown; however, epidemiological studies are currently underway to help answer this question and determine the safe use of phytoestrogens in infants.[65]

References

1. A. L. Ososki and E. J. Kennelly, *Phytother. Res.*, 2003, **17**, 845–869.
2. G. Leclercq, P. de Cremoux, P. This and Y. Jacquot, *Maturitas*, 2011, **68**, 56–64.
3. A. Warri, N. M. Saarinen, S. Makela and L. Hilakivi-Clarke, *Br. J. Cancer*, 2008, **98**, 1485–1493.
4. R. Tsao, Y. Papadopoulos, R. Yang, J. C. Young and K. McRae, *J. Agric. Food Chem.*, 2006, **54**, 5797–5805.
5. R. Maul and S. E. Kulling, *Br. J. Nutr.*, 2010, **103**, 1569–1572.
6. H. Hearnshaw, J. M. Brown, I. A. Cumming, J. R. Goding and M. Nairn, *J. Reprod. Fertil.*, 1972, **28**, 160–161.
7. H. W. Bennetts, E. J. Underwood and F. L. Shier, *Br. Vet. J.*, 1946, **102**, 348–352.
8. N. L. Booth, C. R. Overk, P. Yao, S. Totura, Y. Deng, A. S. Hedayat, J. L. Bolton, G. F. Pauli and N. R. Farnsworth, *J. Agric. Food Chem.*, 2006, **54**, 1277–1282.
9. T. Sabudak and N. Guler, *Phytother. Res.*, 2009, **23**, 439–446.
10. J. E. Rossouw, G. L. Anderson, R. L. Prentice, A. Z. LaCroix, C. Kooperberg, M. L. Stefanick, R. D. Jackson, S. A. Beresford, B. V. Howard, K. C. Johnson, J. M. Kotchen and J. Ockene, *J. Am. Med. Assoc.*, 2002, **288**, 321–333.
11. Y. Du, M. Doren, H. U. Melchert, C. Scheidt-Nave and H. Knopf, *BMC Women's Health*, 2007, **7**, 19.
12. T. L. Dog, R. Marles, G. Mahady, P. Gardiner, R. Ko, J. Barnes, M. L. Chavez, J. Griffiths, G. Giancaspro and N. D. Sarma, *Maturitas*, 2010, **66**, 355–362.
13. S. M. Heinonen, K. Wahala and H. Adlercreutz, *J. Agric. Food Chem.*, 2004, **52**, 6802–6809.

14. N. G. Coldham, C. Darby, M. Hows, L. J. King, A. Q. Zhang and M. J. Sauer, *Xenobiotica*, 2002, **32**, 45–62.

15. J. W. Lampe and J. L. Chang, *Semin. Cancer Biol.*, 2007, **17**, 347–353.

16. J. Howes, M. Waring, L. Huang and L. G. Howes, *J. Altern. Complement. Med.*, 2002, **8**, 135–142.

17. A. Braune, R. Maul, N. H. Schebb, S. E. Kulling and M. Blaut, *Mol. Nutr. Food Res.*, 2010, **54**, 929–938.

18. G. G. Kuiper, J. G. Lemmen, B. Carlsson, J. C. Corton, S. H. Safe, P. T. van der Saag, B. van der Burg and J. A. Gustafsson, *Endocrinology*, 1998, **139**, 4252–4263.

19. A. Pfitscher, E. Reiter and A. Jungbauer, *J. Steroid Biochem. Mol. Biol.*, 2008, **112**, 87–94.

20. K. A. Power and L. U. Thompson, *Breast Cancer Res. Treat.*, 2003, **81**, 209–221.

21. C. Y. Hsieh, R. C. Santell, S. Z. Haslam and W. G. Helferich, *Cancer Res.*, 1998, **58**, 3833–3838.

22. N. Sathyamoorthy and T. T. Wang, *Eur. J. Cancer*, 1997, **33**, 2384–2389.

23. Y. H. Ju, J. Fultz, K. F. Allred, D. R. Doerge and W. G. Helferich, *Carcinogenesis*, 2006, **27**, 856–863.

24. J. T. Hsu, H. C. Hung, C. J. Chen, W. L. Hsu and C. Ying, *J. Nutr. Biochem.*, 1999, **10**, 510–517.

25. T. G. Peterson, G. P. Ji, M. Kirk, L. Coward, C. N. Falany and S. Barnes, *Am. J. Clin. Nutr.*, 1998, **68**, 1505S–1511S.

26. Z. N. Ji, W. Y. Zhao, G. R. Liao, R. C. Choi, C. K. Lo, T. T. Dong and K. W. Tsim, *Gynecol. Endocrinol.*, 2006, **22**, 578–584.

27. C. R. Overk, P. Yao, L. R. Chadwick, D. Nikolic, Y. Sun, M. A. Cuendet, Y. Deng, A. S. Hedayat, G. F. Pauli, N. R. Farnsworth, R. B. van Breemen and J. L. Bolton, *J. Agric. Food Chem.*, 2005, **53**, 6246–6253.

28. D. Grady, T. Gebretsadik, K. Kerlikowske, V. Ernster and D. Petitti, *Obstet. Gynecol.*, 1995, **85**, 304–313.

29. K. H. Humphries and S. Gill, *Can. Med. Assoc. J.*, 2003, **168**, 1001–1010.

30. J. E. Burdette, J. Liu, D. Lantvit, E. Lim, N. Booth, K. P. Bhat, S. Hedayat, R. B. Van Breemen, A. I. Constantinou, J. M. Pezzuto, N. R. Farnsworth and J. L. Bolton, *J. Nutr.*, 2002, **132**, 27–30.

31. F. Occhiuto, R. D. Pasquale, G. Guglielmo, D. R. Palumbo, G. Zangla, S. Samperi, A. Renzo and C. Circosta, *Phytother. Res.*, 2007, **21**, 130–134.

32. D. L. Alves, S. M. Lima, C. R. da Silva, M. A. Galvao, A. Shanaider, R. A. de Almeida Prado and T. Aoki, *Maturitas*, 2008, **61**, 364–370.

33. S. Kawakita, F. Marotta, Y. Naito, U. Gumaste, S. Jain, J. Tsuchiya and E. Minelli, *Clin. Interv. Aging*, 2009, **4**, 91–100.

34. S. A. Atkinson and W. E. Ward, *Can. Med. Assoc. J.*, 2001, **165**, 1511–1514.

35. B. L. Clarke and S. Khosla, *Arch. Biochem. Biophys.*, 2010, **503**, 118–128.

36. E. Reiter, V. Beck, S. Medjakovic, M. Mueller and A. Jungbauer, *Menopause*, 2009, **16**, 1049–1060.

37. H. Ha, H. Y. Lee, J. H. Lee, D. Jung, J. Choi, K. Y. Song, H. J. Jung, J. S. Choi, S. I. Chang and C. Kim, *Arch. Pharm. Res.*, 2010, **33**, 625–632.

38. S. E. Geller, L. P. Shulman, R. B. van Breemen, S. Banuvar, Y. Zhou, G. Epstein, S. Hedayat, D. Nikolic, E. C. Krause, C. E. Piersen, J. L. Bolton, G. F. Pauli and N. R. Farnsworth, *Menopause*, 2009, **16**, 1156–1166.

39. S. E. Geller and L. Studee, *Climacteric*, 2006, **9**, 245–263.

40. M. M. Terzic, J. Dotlic, S. Maricic, T. Mihailovic and B. Tosic-Race, *J. Obstet. Gynaecol. Res.*, 2009, **35**, 1091–1095.

41. J. T. Coon, M. H. Pittler and E. Ernst, *Phytomedicine*, 2007, **14**, 153–159.

42. J. B. Howes, K. Bray, L. Lorenz, P. Smerdely and L. G. Howes, *Climacteric*, 2004, **7**, 70–77.

43. M. Lipovac, P. Chedraui, C. Gruenhut, A. Gocan, M. Stammler and M. Imhof, *Maturitas*, 2010, **65**, 258–261.

44. E. Reiter, V. Beck, S. Medjakovic, M. Mueller and A. Jungbauer, *Menopause*, 2009, **16**, 1049–1060.

45. V. Beck, U. Rohr and A. Jungbauer, *J. Steroid Biochem. Mol. Biol.*, 2005, **94**, 499–518.

46. N. L. Booth, C. R. Overk, P. Yao, J. E. Burdette, D. Nikolic, S. N. Chen, J. L. Bolton, R. B. van Breemen, G. F. Pauli and N. R. Farnsworth, *J. Altern. Complement. Med.*, 2006, **12**, 133–139.

47. E. E. Krebs, K. E. Ensrud, R. MacDonald and T. J. Wilt, *Obstet. Gynecol.*, 2004, **104**, 824–836.

48. A. E. Lethaby, J. Brown, J. Marjoribanks, F. Kronenberg, H. Roberts and J. Eden, *Cochrane Database Syst. Rev.*, 2007, CD001395.

49. H. D. Nelson, K. K. Vesco, E. Haney, R. Fu, A. Nedrow, J. Miller, C. Nicolaidis, M. Walker and L. Humphrey, *J. Am. Med. Assoc.*, 2006, **295**, 2057–2071.

50. P. H. van de Weijer and R. Barentsen, *Maturitas*, 2002, **42**, 187–193.

51. L. A. Hidalgo, P. A. Chedraui, N. Morocho, S. Ross and G. San Miguel, *Gynecol. Endocrinol.*, 2005, **21**, 257–264.

52. M. Imhof, A. Gocan, F. Reithmayr, M. Lipovac, C. Schimitzek, P. Chedraui and J. Huber, *Maturitas*, 2006, **55**, 76–81.

53. N. L. Booth, C. E. Piersen, S. Banuvar, S. E. Geller, L. P. Shulman and N. R. Farnsworth, *Menopause*, 2006, **13**, 251–264.

54. G. E. Hale, C. L. Hughes, S. J. Robboy, S. K. Agarwal and M. Bievre, *Menopause*, 2001, **8**, 338–346.

55. P. B. Clifton-Bligh, R. J. Baber, G. R. Fulcher, M. L. Nery and T. Moreton, *Menopause*, 2001, **8**, 259–265.

56. L. Hooper, G. Madhavan, J. A. Tice, S. J. Leinster and A. Cassidy, *Human Reprod. Update*, 2010, **16**, 745–760.

57. T. J. Powles, A. Howell, D. G. Evans, E. V. McCloskey, S. Ashley, R. Greenhalgh, J. Affen, L. A. Flook and A. Tidy, *Menopause Int.*, 2008, **14**, 6–12.

58. C. Atkinson, R. M. Warren, E. Sala, M. Dowsett, A. M. Dunning, C. S. Healey, S. Runswick, N. E. Day and S. A. Bingham, *Breast Cancer Res.*, 2004, **6**, R170–R179.

59. C. M. Weaver, B. R. Martin, G. S. Jackson, G. P. McCabe, J. R. Nolan, L. D. McCabe, S. Barnes, S. Reinwald, M. E. Boris and M. Peacock, *J. Clin. Endocrinol. Metab.*, 2009, **94**, 3798–3805.

60. C. Atkinson, J. E. Compston, N. E. Day, M. Dowsett and S. A. Bingham, *Am. J. Clin. Nutr.*, 2004, **79**, 326–333.

61. K. D. Setchell, L. Zimmer-Nechemias, J. Cai and J. E. Heubi, *Lancet*, 1997, **350**, 23–27.

62. W. N. Jefferson and C. J. Williams, *Reprod. Toxicol.*, 2011, **31**, 272–279.

63. Y. A. Cao, A. M. Calafat, D. R. Doerge, D. M. Umbach, J. C. Bernbaum, N. C. Twaddle, X. I. Ye and W. J. Rogan, *J. Expo. Sci. Environ. Epidemiol.*, 2009, **19**, 223–234.

64. L. Hoey, I. R. Rowland, A. S. Lloyd, D. B. Clarke and H. Wiseman, *Br. J. Nutr.*, 2004, **91**, 607–616.

65. T. M. Badger, J. M. Gilchrist, R. T. Pivik, A. Andres, K. Shankar, J.-R. Chen and M. J. Ronis, *Am. J. Clin. Nutr.*, 2009, **89**, 1668S–1672S.

66. W. N. Jefferson, E. Padilla-Banks and R. R. Newbold, *Mol. Nutr. Food Res.*, 2007, **51**, 832–844.

67. W. N. Jefferson, E. Padilla-Banks and R. R. Newbold, *J. AOAC Int.*, 2006, **89**, 1189–1196.

68. W. N. Jefferson, D. Doerge, E. Padilla-Banks, K. A. Woodling, G. E. Kissling and R. Newbold, *Environ. Health Perspect.*, 2009, **117**, 1883–1889.

69. M. A. Cimafranca, J. Davila, G. C. Ekman, R. N. Andrews, S. L. Neese, J. Peretz, K. A. Woodling, W. G. Helferich, J. Sarkar, J. A. Flaws, S. L. Schantz, D. R. Doerge and P. S. Cooke, *Biol. Reprod.*, 2010, **83**, 114–121.

70. K. D. Setchell, L. Zimmer-Nechemias, J. Cai and J. E. Heubi, *Am. J. Clin. Nutr.*, 1998, **68**, 1453S–1461S.

71. A. A. D'Aloisio, D. D. Baird, L. A. DeRoo and D. P. Sandler, *Environ. Health Perspect.*, 2010, **118**, 375–381.

72. J. C. Bernbaum, D. M. Umbach, N. B. Ragan, J. L. Ballard, J. I. Archer, H. Schmidt-Davis and W. J. Rogan, *Environ. Health Perspect.*, 2008, **116**, 416–420.

73. J. Kaludjerovic and W. E. Ward, *J. Nutr.*, 2009, **139**, 467–473.

74. W. E. Ward, S. Kim, D. Chan and D. Fonseca, *J. Nutr. Biochem.*, 2005, **16**, 743–749.

75. C. M. Weaver and L. L. Legette, *J. Nutr.*, 2010, **140**, 1377S–1379S.

76. J. Kaludjerovic and W. E. Ward, *J. Nutr.*, 2010, **140**, 766–772.

77. J. F. Stevens, A. W. Taylor, J. E. Clawson and M. L. Deinzer, *J. Agric. Food Chem.*, 1999, **47**, 2421–2428.

78. J. F. Stevens, A. W. Taylor and M. L. Deinzer, *J. Chromatogr. A*, 1999, **832**, 97–107.

79. S. R. Milligan, J. C. Kalita, V. Pocock, V. Van De Kauter, J. F. Stevens, M. L. Deinzer, H. Rong and D. De Keukeleire, *J. Clin. Endocrinol. Metab.*, 2000, **85**, 4912–4915.

80. N. G. Coldham and M. J. Sauer, *Food Chem. Toxicol.*, 2001, **39**, 1211–1224.

81. A. Heyerick, S. Vervarcke, H. Depypere, M. Bracke and D. De Keukeleire, *Maturitas*, 2006, **54**, 164–175.

82. G. Morali, F. Polatti, E. N. Metelitsa, P. Mascarucci, P. Magnani and G. B. Marre, *Arzneim.-Forsch.*, 2006, **56**, 230–238.

83. E. Brunelli, A. Minassi, G. Appendino and L. Moro, *J. Steroid Biochem. Mol. Biol.*, 2007, **107**, 140–148.

84. M. Verzele, *J. Inst. Brewing*, 1986, **92**, 32–48.

85. S. R. Milligan, J. C. Kalita, A. Heyerick, H. Rong, L. De Cooman and D. De Keukeleire, *J Clin. Endocrinol. Metab.*, 1999, **84**, 2249–2252.

86. D. Intelmann, G. Haseleu and T. Hofmann, *J. Agric. Food Chem.*, 2009, **57**, 1172–1182.

87. J. Tekel, D. De Keukeleire, H. Rong, E. Daeseleire and C. Van Peteghem, *J. Agric. Food Chem.*, 1999, **47**, 5059–5063.

88. T. Nomura, Y. Hano and T. Fukai, *Proc. Jpn. Acad. Ser. B, Phys. Biol. Sci.*, 2009, **85**, 391–408.

89. M. Kitaoka, H. Kadokawa, M. Sugano, K. Ichikawa, M. Taki, S. Takaishi, Y. Iijima, S. Tsutsumi, M. Boriboon and T. Akiyama, *Planta Med.*, 1998, **64**, 511–515.

90. M. Miyamoto, Y. Matsushita, A. Kiyokawa, C. Fukuda, Y. Iijima, M. Sugano and T. Akiyama, *Planta Med.*, 1998, **64**, 516–519.

91. F. Z. Bohlmann, C. King, R. M. Robinson and H. Robinson, *Phytochemistry*, 1979, **18**, 1246–1247.

92. M. Komatsu, I. Yokoe and Y. Shirataki, *Chem. Pharm. Bull.*, 1978, **26**, 3863–3870.

93. S. V. Tokalov, Y. Henker, P. Schwab, P. Metz and H. O. Gutzeit, *Pharmacology*, 2004, **71**, 46–56.

94. C. R. Overk, P. Yao, L. R. Chadwick, D. Nikolic, Y. Sun, M. A. Cuendet, Y. Deng, A. S. Hedayat, G. F. Pauli, N. R. Farnsworth, R. B. van Breemen and J. L. Bolton, *J. Agric. Food Chem.*, 2005, **53**, 6246–6253.

95. O. Zierau, J. Hamann, S. Tischer, P. Schwab, P. Metz, G. Vollmer, H. O. Gutzeit and S. Scholz, *Biochem. Biophys. Res. Commun.*, 2005, **326**, 909–916.

96. D. A. Amer, G. Kretzschmar, N. Muller, N. Stanke, D. Lindemann and G. Vollmer, *J. Steroid Biochem. Mol. Biol.*, 2010, **120**, 208–217.

97. D. Nikolic, Y. Li, L. R. Chadwick, G. F. Pauli and R. B. van Breemen, *J. Mass Spectrom.*, 2005, **40**, 289–299.

98. Y. Pang, D. Nikolic, D. Zhu, L. R. Chadwick, G. F. Pauli, N. R. Farnsworth and R. B. van Breemen, *Mol. Nutr. Food Res.*, 2007, **51**, 872–879.

99. J. Guo, D. Nikolic, L. R. Chadwick, G. F. Pauli and R. B. van Breemen, *Drug Metab. Dispos.*, 2006, **34**, 1152–1159.

100. D. Nikolic, Y. Li, L. R. Chadwick, S. Grubjesic, P. Schwab, P. Metz and R. B. van Breemen, *Drug Metab. Dispos.*, 2004, **32**, 272–279.

101. S. Possemiers, A. Heyerick, V. Robbens, D. De Keukeleire and W. Verstraete, *J. Agric. Food Chem.*, 2005, **53**, 6281–6288.

102. O. Zierau, S. Hauswald, P. Schwab, P. Metz and G. Vollmer, *J. Steroid Biochem. Mol. Biol.*, 2004, **92**, 107–110.

103. G. Kretzschmar, O. Zierau, J. Wober, S. Tischer, P. Metz and G. Vollmer, *J. Steroid Biochem. Mol. Biol.*, 2010, **118**, 1–6.

104. O. Zierau, C. Morrissey, R. W. Watson, P. Schwab, S. Kolba, P. Metz and G. Vollmer, *Planta Med.*, 2003, **69**, 856–858.

105. P. Diel, R. B. Thomae, A. Caldarelli, O. Zierau, S. Kolba, S. Schmidt, P. Schwab, P. Metz and G. Vollmer, *Planta Med.*, 2004, **70**, 39–44.

106. O. Zierau, G. Kretzschmar, F. Moller, C. Weigt and G. Vollmer, *Mol. Cell. Endocrinol.*, 2008, **294**, 92–99.

107. C. R. Overk, J. Guo, L. R. Chadwick, D. D. Lantvit, A. Minassi, G. Appendino, S. N. Chen, D. C. Lankin, N. R. Farnsworth, G. F. Pauli, R. B. van Breemen and J. L. Bolton, *Chem. Biol. Interact.*, 2008, **176**, 30–39.

108. M. Bottner, J. Christoffel and W. Wuttke, *J. Endocrinol.*, 2008, **198**, 395–401.

109. G. Rimoldi, J. Christoffel and W. Wuttke, *Menopause*, 2006, **13**, 669–677.

110. M. Humpel, P. Isaksson, O. Schaefer, U. Kaufmann, P. Ciana, A. Maggi and W. D. Schleuning, *J. Steroid Biochem. Mol. Biol.*, 2005, **97**, 299–305.

111. S. Bolca, J. Li, D. Nikolic, N. Roche, P. Blondeel, S. Possemiers, D. De Keukeleire, M. Bracke, A. Heyerick, R. B. van Breemen and H. Depypere, *Mol. Nutr. Food Res.*, 2010, **54** (suppl 2), S284–S294.

112. R. R. Newbold, E. P. Banks, B. Bullock and W. N. Jefferson, *Cancer Res.*, 2001, **61**, 4325–4328.

113. W. N. Jefferson, J. F. Couse, E. Padilla-Banks, K. S. Korach and R. R. Newbold, *Biol. Reprod.*, 2002, **67**, 1285–1296.

114. W. N. Jefferson, E. Padilla-Banks and R. R. Newbold, *Biol. Reprod.*, 2005, **73**, 798–806.

115. W. Jefferson, R. Newbold, E. Padilla-Banks and M. Pepling, *Biol. Reprod.*, 2006, **74**, 161–168.

Role of Metabolism in the Bioactivation/Detoxification of Food Contaminants

JEAN-PIERRE CRAVEDI AND DANIEL ZALKO

UMR 1331 TOXALIM, INRA, 180 chemin de Tournefeuille, BP 93173, F-31027 Toulouse Cédex 3, France

6.1 Introduction

A number of potentially toxic contaminants are found regularly in foods. Some of them are well-known environmental contaminants such as dioxins, heavy metals or agrochemical residues. Others are natural compounds (mycotoxins, phycotoxins) or are produced during processing. For instance, 2-amino-1-methyl-6-phenylimidazo[4,5-*b*]pyridine (PhIP) and benzo[*a*]pyrene are aromatic compounds formed during broiling/frying of meat and the metabolites of these chemicals have been detected in human urine. Furthermore, when present in food at a high level, some micronutrients may also have detrimental effects on health, in particular when exposure occurs during early life stages.

The toxicity of a food contaminant is not only determined by its intrinsic reactivity and by the administered dose. The bioavailability of the compound for the target organ or tissue is an important factor to take into account in the evaluation of the toxic response, but also in assessing the potential for bioaccumulation. In mammals, a xenobiotic that is ingested, and later absorbed along the gastrointestinal tract, is transported through the portal circulation to the liver, where it is often subjected to hepatic metabolism, followed by

Issues in Toxicology No. 11
Hormone-Disruptive Chemical Contaminants in Food
Edited by Ingemar Pongratz and Linda Vikström Bergander

Published by the Royal Society of Chemistry, www.rsc.org

elimination into the bile or in urine. Although occurring at lower levels compared to the liver, direct intestinal metabolism can also take place during the absorption process. The biotransformation of the parent compound results in a significant reduction of the unmetabolized fraction available for the biological targets. This phenomenon is known as the first-pass effect.

The biotransformation process comprises a series of biochemical reactions by which the structure of a compound is changed during passage through the organisms, facilitating its excretion from the body. A typical xenobiotic metabolism pathway involves the oxidation of the parent compound (phase I oxidation), followed by conjugation of the oxidized moiety with highly polar endogenous molecules such as glucuronic acid, sulfate, glutathione or amino acids (phase II reactions). Enzymes that catalyze phase I reactions are mainly located in the endoplasmic reticulum and for the most part are cytochrome P450-dependent monooxygenases. Cytochrome P450s (CYPs) are a superfamily of heme-containing enzymes involved in oxidative, reductive and peroxidative biotransformation of many endogenous (*e.g.* steroids, bile acids, fatty acids, prostaglandins) and exogenous (plant products, drugs, environment pollutants) compounds.[1] The general mode of action is the incorporation of one oxygen atom into the substrate by a one-electron transfer mechanism involving NADPH as reducing cofactor.[2] Of the 57 human CYPs, 15 are predominantly involved in the biotransformation of xenobiotics, among which CYP3A4 is the major contributor.[3] Other phase I biotransformation enzymes include flavin-dependent monooxygenase, peroxidase, epoxide hydrolase, carboxylesterase, amidase, alcohol dehydrogenase and monoamine oxidase. An important feature associated with CYPs is their induction potential. CYPs are tightly regulated by many substrates, including xenobiotics. For most CYPs, the regulation is transcriptional; for example, CYP1A1, CYP1A2 and CYP1B1 are regulated *via* the aryl hydrocarbon receptor (AhR), while others such as CYP3A are regulated by the pregnane X receptor (PXR) and/or the constitutively active receptor (CAR). Nevertheless, translational and posttranslational regulation also occurs. The result is that enhanced expression of P450 activity can greatly augment the detoxification/bioactivation rates of potentially toxic compounds.[4] Genetic polymorphisms have been reported for several CYPs, in particular 2C9, 2C19 and 2D6, and can result in reduced or increased enzyme activity in specific subpopulations.[5]

The most important phase II reactions, in terms of the quantities and number of xenobiotics metabolized, are glucuronide conjugation, sulfate conjugation and glutathione conjugation. Whereas the substrates for the glucuronidation and sulfate conjugation are nucleophiles, the substrates for glutathione conjugation are electrophiles. Uridine-diphosphate glucuronosyl transferase (UGT) is an enzyme superfamily that catalyzes the transfer of a glucuronic acid moiety (from uridine-diphosphate glucuronic acid, UDPGA) to the acceptor substrate. UGTs are located in the membrane of the endoplasmic reticulum, whereas UDPGA is synthesized in the cytosol from glucose, then actively transported into the endoplasmic reticulum. More than 20 UGTs have been identified in humans.[6] Many of the substrates that undergo glucuronidation are also metabolized by sulfotransferases (SULTs) with phosphoadenosine-phosphosulfate (PAPS) as co-substrate. There are nine genes encoding cytosolic SULTs in humans and they belong to the SULT1 and SULT2 families. One important feature that differentiates SULTs from other enzymes is that, once formed, the sulfuric acid esters can be hydrolyzed, followed by re-sulfation and continued cycling, resulting in an increase in the overall exposure.[4] In addition, sulfate conjugation and glucuronidation often play an important role in the process of enterohepatic circulation. These conjugates may be hydrolyzed by gut bacteria that free the xenobiotic, which consequently can be reabsorbed by the intestine and transported to the liver. At this stage, the conjugate may be reformed, resulting in a significant increase of the residence time of the xenobiotic in the body. For example, research projects conducted in animals have demonstrated that diethylstilbestrol (DES) is almost totally excreted in the bile, in rodents. The blockage of this pathway (by bile duct cannulation) results in a pronounced increase in the residence time of DES in the body and in an increase in the toxicity of DES by 24-fold in mice and 130-fold in rat.[7]

GSTs are homo- or heterodimers composed of relatively small proteins that are primarily expressed as soluble cytosolic proteins. A small proportion, however, is expressed as a membrane-bound enzyme and is consequently found in microsomes. In humans, the complete GST gene family comprises 16 genes in six subfamilies (alpha, mu, omega, pi, theta, zeta). GST-kappa 1 (GSTK1), prostaglandin E synthase and three microsomal GSTs (MGST1, MGST2, MGST3) have been shown to encode the membrane-bound enzymes having a GST-like activity, demonstrating that these genes are not evolutionarily related to the GST gene family.[8] The most abundant mammalian GST enzymes belong to alpha, mu and pi classes. They catalyze the conjugation of a number of electrophilic compounds with reduced glutathione. Glutathione conjugates can be excreted as such, in bile, or can be converted to mercapturic acids in the kidney and excreted in the urine. As observed for CYPs, polymorphisms of conjugating enzymes in humans exist for most of the isoforms. Moreover, induction of GSTs and UGTs may occur at a significant extent, although not comparable with many inductions of the P450 system.

Biotransformations generally result in the detoxification of the parent xenobiotic through the formation of harmless metabolites and the facilitation of residue excretion. In some instances, however, metabolism can result in the formation of molecules which are biologically more active than the parent compound towards cellular targets, among which are cell receptors. Such steps can potentially trigger an impairment of physiological regulations. Many of these activation reactions are oxidations that are catalyzed by the cytochrome P450 class of monooxygenases. For example, CYP1A1, CYP1A2, CYP2A6, CYP2AE1 and CYP3A4 are involved in the activation of benzo-[*a*]pyrene, PhIP, dimethylnitrosamine, ethyl carbamate and aflatoxin B1, respectively, leading to genotoxic intermediates and explaining the carcinogenicity of these compounds.[5] However, the range of potential activating reactions is large, and cytosolic enzymes and non-oxidative pathways may also play a role in the generation of active metabolites. UGTs, SULTs and GSTs may catalyze the bioactivation of xenobiotics.[9] For instance, the sulfation of *N*-(hydroxyacetyl)amidofluorene, a metabolite of *N*-acetylamido-fluorene, results in the formation of a highly reactive *O*-sulfate ester which is acid-labile and decomposes to a nitrenium ion, causing DNA damage. The same type of phenomenon occurs with 2-naphthylamine after conjugation with glucuronic acid.[10] Reactive metabolites are also produced from glu-tathione conjugates.[11] For example, glutathione conjugates of pesticides such as propachlor and chlorothalonil are broken down by γ-glutamyltransferase and dipeptidase to cysteine conjugates which are further acetylated to give mercapturic acids or cleaved by β-lyase enzymes to give reactive intermediates.[12,13]

Although extensive studies have resulted in relatively detailed information on the metabolic activation of procarcinogens to ultimate carcinogens, only a limited number of investigators has focused on the bioactivation of endocrine disruptors. Recently, a review covering environmental proestrogens that are activated by hepatic xenobiotic metabolizing enzymes has been published.[14] Nevertheless, today it is well known that endocrine disruption is not limited to the interaction of chemicals with estrogen receptors. The mechanisms are much broader than originally recognized: endocrine disruptors can act *via* several hormone receptors and transporters, but also on enzymatic pathways involved in steroid biosynthesis and/or metabolism.[15] The focus of this chapter is on the biotransformation of selected endocrine disruptors. Since the list of molecules identified is highly heterogeneous and is growing every month, it would be difficult to review the metabolism of all known endocrine disruptors. Therefore, our choice is preferentially on food contaminants that are not directly endocrine disruptors (or weak endocrine disruptors), but which activate through biotransformation processes. Our efforts have been concentrated on rodent and human data. Various categories of compounds have been selected, including pesticides, plasticizers, flame retardants, ink components and natural plant products.

6.2 Equol

One of the most active isoflavone metabolites in mammals is equol. This metabolite has been reported to be present as free or as a glucuronide/sulfate conjugate in urine and/or plasma of many animal species, including ruminants, hens, monkeys, dogs, rodents and pigs.[16] In addition, equol has been identified in various milk samples.[17,18] All animal species investigated have been reported to produce equol in response to the consumption of isoflavones, whether from soy protein or clover. Studies carried out in humans fed soy-based foods sug-gest that only part of the population produce equol. In western countries, approximately 25–30% of soy food consumers produce equol, whereas the frequency in Asia is in the 50–60% range.[19] The reasons for these differences are still unclear, but it is likely that both interindividual differences in isoflavone pharmacokinetics and the type of soy foods consumed play a major role in the observed variations.[20–23] One parameter of particular interest is the ratio between daidzein and its glycoside precursor daidzin in these foods.[24]

Isoflavones in soy foods are conjugated to sugars to form β-glycosides. The β-glycosides require hydrolysis prior intestinal absorption.[25] This metabolic step occurs along the entire length of the intestine and involves both the brush border membrane and microflora β-glycosidases.[26] The aglycone daidzein is released from the hydrolysis of the β-glycoside daizin and further metabolism takes place (Figure 6.1). After absorption, in the small intestine, daidzein is conjugated with glucuronic acid and sulfate by hepatic phase II enzymes (UGTs and SULTs), or can be hydroxylated to form catechols such as 3′-OH-daidzein, 6-OH-daidzein and 8-OH-daidzein.[27,28] Like endogenous estrogens, these conjugates are excreted both through urine and bile. After excretion into bile, conjugated daidzein can be deconjugated by gut bacteria and further metabolized. The conversion of daidzein to equol is due to the action of bacterial flora and occurs *via* a reductive pathway that involves the formation of the intermediate dihydrodaidzein.[25,29] Dihydrodaidzein also produces demethylangolensin (DMA). Once formed, equol can undergo both phase I[30] and phase II reactions.[31]

The daidzein metabolites equol and DMA are found to be more potent estrogens than their parent compound.[32–34] Equol possesses high estrogenic

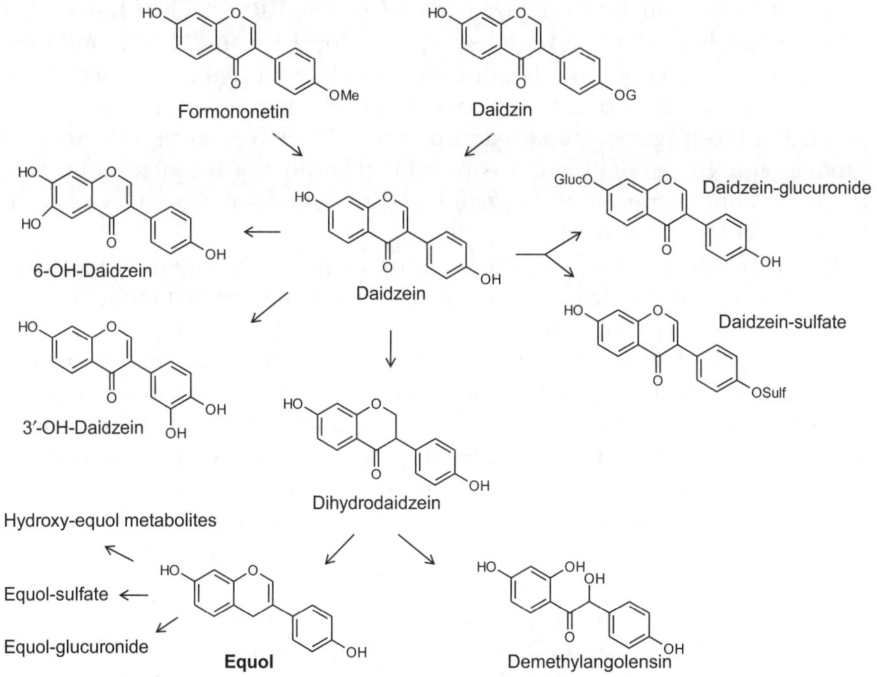

Figure 6.1   Synthesis and biotransformation of daidzein in mammals (G = glucose, Sulf = sulfate).

activities: the binding affinities to estrogen receptors (ER) α and β are higher than those of daidzein and similar to those of genistein, with a preference for ERβ. Furthermore, equol induces binding of the activated receptor to the estrogen response element, as well as transcription, more strongly than other isoflavones, especially regarding ERα.[35–37] In MCF7 cells, equol and DMA have been found to be more potent than daidzein in stimulating cell growth.[35,38] Equol was also found to be approximately 100-fold more potent than daidzein in stimulating pS2 (an estrogen inducible protein) mRNA expression.[39] Schmitt *et al.* compared directly the activities of DMA with those of equol and found that the latter exhibited a higher binding affinity than DMA towards ERα.[40] As a result of a chiral carbon at position C3 of the molecule, equol exists in two enantiomeric forms: (*R*)-(+)-equol and (*S*)-(−)-equol. The latter is the natural diastereoisomer produced by human and rat intestinal bacteria.[19] The binding affinities of (*R*)- and (*S*)-equol, as well as the racemate, towards ERα and ERβ have been investigated, showing that (*S*)-equol exhibits a high binding affinity for ERβ, whereas (*R*)-equol binds less strongly and with a preference for ERα.[37] Equol was also found to bind to sex hormone binding globulin (SHBG) and to competitively inhibit estradiol and testosterone binding in a dose-dependent manner.[41] Moreover, equol can act as an anti-androgen by binding 5α-dihydrotestosterone with high affinity, which prevents 5α-dihydrotestosterone from binding the androgen receptor.[42] Mueller and co-workers tested the agonist and antagonist properties of 3′-OH-daidzein and 6-OH-daidzein on ERα and ERβ.[43] They found that these metabolites showed very low affinity for ERα and ERβ, but were considered as superagonists because they exhibited a higher fold induction than diethylstilbestrol or estradiol. Some *in vivo* experiments have confirmed that equol is estrogenic and anti-androgenic.[34] Moreover, given that animals produce equol from daidzein, it is possible that most of the effects observed following administration of daidzein to experimental animals solely resulted from the biological properties of equol.

The metabolic pathways of daidzein are of particular interest since they show that, on the one hand, the formation of conjugated metabolites corresponds to an inactivation process, but, on the other hand, equol (and to a lesser extent DMA) are more potent estrogenic compounds than the parent isoflavone. In addition, given the fact that daidzein is mainly present in food as a glycoside (daidzin), which is neither bioavailable nor biologically active, its hydrolysis may also be considered as a first step in the bioactivation process. In the case of isoflavones, it must be noted that some compounds have been reported to exhibit genotoxic potential in cultured cells.[44] Lehman *et al.* have found that equol and 3′-OH-daidzein exhibit a clastogenic potential and induce micronuclei and DNA strand breaks, which did not correlate, in these studies, with their estrogenic potential.[45] These data indicate that the production of equol has to be considered as a bioactivation pathway, not only based on ERβ activity, which is usually considered as a beneficial effect, but also because this metabolite has been demonstrated to be genotoxic *in vitro*.

6.3 Polychlorinated Biphenyls

Polychlorinated biphenyls (PCBs) were commercially produced as complex mixtures containing multiple isomers with a different degree of chlorination for a variety of applications, including dielectric fluids for capacitors and transformers. PCBs have entered the environment both through use and disposal. Since PCBs do not readily degrade in the environment and are lipophilic, they accumulate in animals and humans and biomagnify in the food chain. The main pathway of human exposure for the majority of the population is *via* food consumption, with the exception of specific cases of accidental or occupational exposure. The intake estimate for the sum of PCBs (most of which are non-dioxin-like PCBs) in the general adult population is approximately in the 5–12 ng kg^{-1} bw range.[46,47] Sirot *et al.* estimated this intake in French high seafood consumers to be 57 ng kg^{-1} bw, whereas EFSA estimated the median daily intake for an exclusively breastfed infant as 1.5 µg kg^{-1} bw.[48,49]

PCBs are extensively absorbed from the gastrointestinal tract and mainly distributed in lipid-rich tissues. Transfer across cell membranes is by passive diffusion, and PCBs as well as their corresponding hydroxylated metabolites are transferred to the fetus.[50–52]

A general scheme for the metabolism of PCBs is shown in Figure 6.2. The first step is the introduction of an oxygen atom *via* the CYP enzyme system. Oxygen may be added across an aromatic ring bond to give an arene oxide, or directly inserted to give a hydroxylated metabolite. The hydroxylated

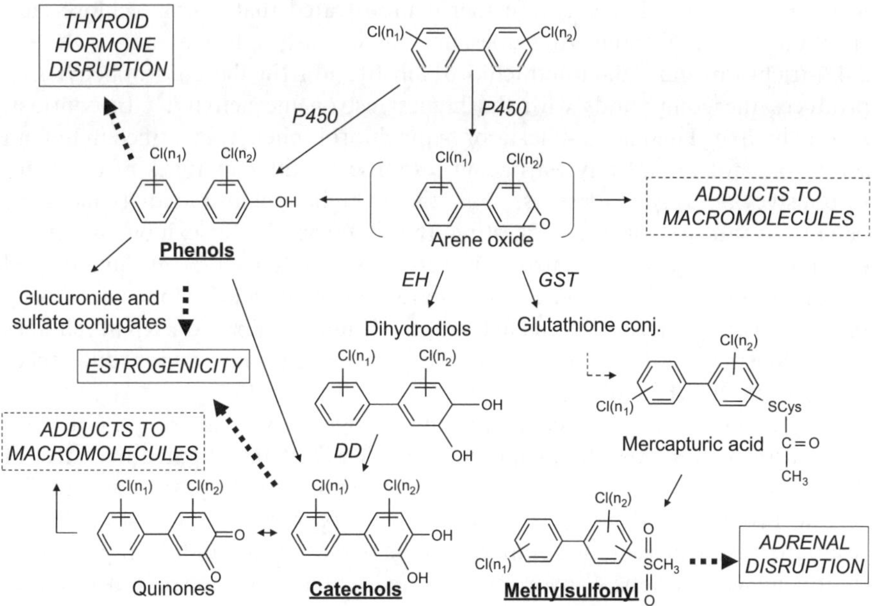

Figure 6.2 *In vitro/in vivo* metabolic pathways of PCBs and the biological effects of reactive metabolites.

metabolite may further be conjugated with glucuronide or sulfate in reactions catalyzed by UGTs or SULTs, respectively.[53] Further hydroxylation of hydroxylated metabolites is also possible. If the initial metabolite formed by CYP is an arene oxide, it can rearrange to a phenolic hydroxyl group, be further metabolized by epoxide hydrolase to form a dihydrodiol, or undergo conjugation with glutathione in a reaction catalyzed by GST. The dihydrodiol may also be oxidized to form a catechol. If a glutathione conjugate is formed from the epoxide, it may ultimately be converted to a mercapturic acid, or enter an alternative pathway leading to the formation of a methyl sulfone metabolite. Arene oxides of PCBs are reactive electrophilic intermediates that can form adducts with DNA and proteins. Catechol metabolites are in equilibrium with their oxidized form, the corresponding hydroquinone and quinone.[53] Both are reactive intermediates with the potential for adduct formation.[54] With respect to the DNA damage process, these reactions can be considered as bioactivation pathways, whereas production of glucuronides, sulfate conjugates and glutathione conjugates are essentially detoxication pathways.

PCBs have also been reported to be reprotoxicants and endocrine disruptors. These effects are very likely triggered by the formation of certain metabolites. Korach and co-workers first showed, in 1998, that several hydroxy-PCBs competitively bind the mouse ER.[55] In their study, two of the most active agonists were 2′,4′,6′-trichlorobiphenyl-4-ol and 2′,3′,4′,5′-tetrachloro-biphenyl-4-ol, suggesting that the most active congeners contained a single *para*-hydroxy group on one of the two phenyl rings. Structure–activity studies carried out with a series of tetra- and pentachloro congeners, as well as various *in vitro* and *in vivo* bioassays, further demonstrated that a single chloro sub-stitution of the phenolic ring does not significantly affect activity, whereas 2,4,6-trichloro and 2,3,4,6-tetrachloro substitution on the chlorophenyl ring produces the compounds with the highest estrogenic activity.[56] In contrast, several hydroxylated penta-, hexa- or heptachlorobiphenyls identified in human serum did not induce any estrogenic response in different types of estrogen-responsive *in vitro* bioassays, but these higher chlorinated biphenylols behave as anti-estrogens, suggesting that hydroxy-PCBs exhibit structure-dependent estrogenic/anti-estrogenic activities.[57,58] Kuiper *et al.* investigated the estrogenic activity of various tri-, tetra- and pentachlorobiphenylols using several *in vitro* assays, including activation of human ERα and ERβ in cell lines transfected with these receptors.[59] They observed only minor differences between the two ER subtypes.

Only a few studies have compared the estrogenicity of hydroxy-PCBs with their corresponding parent compounds. Fielden *et al.* investigated the effect of 2,2′,4,6,6′-pentachlorobiphenyl and the corresponding 4-hydroxy metabolite on the proliferation of MCF7 cells.[60] They observed a 2- to 3-fold higher response for the metabolite compared to the parent compound. Intraperitoneal treatment with various 4′-hydroxytetra- and pentachlorobiphenyl metabolites resulted in a significant increase in uterus weight at the lowest dose tested ($25\,\mathrm{mg\,kg^{-1}\,bw\,d^{-1}}$).[61] In addition to their direct binding and activation of ER, hydroxylated metabolites of some PCBs can also exert estrogenic effects through an entirely different mechanism. Some hydroxylated PCBs (in particular those having two adjacent chlorine substituents around the hydroxyl group) are potent inhibitors of sulfotransferase,[62–64] an enzyme that is critical for the sulfation, and consequently the inactivation, of estradiol.[65]

Some PCB catechol metabolites have been shown to possess similar estrogenic activities as those of monohydroxylated metabolites, suggesting that further oxidative metabolism of estrogenic phenolic metabolites would not necessarily result in lowering the total estrogenic burden of a PCB-exposed organism.[66] The extent of chlorination and the position of the catechol are important factors in the estrogenic response. Whereas 2,3-catechols show no detectable activity in HeLa cell lines transfected with mouse ER cDNA associated with a reporter gene, the activity of 3,4-catechols have been found to increase with the degree of chlorination. Another possible pathway for the estrogenic effect of catechol metabolites could be the inhibition of the catechol-*O*-methyltransferase (COMT).[67,68] PCB catechols are COMT substrates with kinetic properties similar to those measured with catechol metabolites of estradiol. COMT inhibition by PCB catechols is a partial noncompetitive mechanism, resulting in an alteration of the metabolism of catechol estrogens, thus indirectly contributing to PCB-mediated endocrine effects.[68] Since catechols are often reactive metabolites, their formation (and lack of detoxication by COMT) can also contribute to PCB-induced tumorigenesis.

The toxicity of dioxin-like (co-planar) PCBs is thought to be mediated mainly by the aryl hydrocarbon receptor (AhR). Recently, Kamata *et al.* measured the AhR agonist activity of monohydroxylated PCBs and their non-hydroxylated analogs using yeast cells transduced with the human AhR and its response pathway.[69] Of 49 hydroxy-PCBs that had the same chlorination patterns as the tested PCBs, 26 had activities that were more than twice higher than the corresponding PCBs, or were found to be active while the corresponding PCB was not. In particular, 3′,4,5′-trichlorobiphenyl-2-ol and 3′,4,4′-trichlorobiphenyl-3-ol were 37- and 22-fold more potent than their non-hydroxylated analogs and were as active as the standard inducer β-naphthoflavone.

Many studies clearly support the evidence that PCBs can affect circulating levels of thyroid hormones in experimental animals[50,70,71] and in humans.[72] Hydroxy-PCBs are suspected to play a major role in these effects, since metabolites with the hydroxyl group substituted at the *meta*- or *para*-positions and one or more chlorine atoms substituted adjacent to the hydroxyl group on one or both aromatic ring(s) showed high potency for displacing thyroxine from transthyretin.[53,71,73] Moreover, some hydroxylated metabolites have been shown to be potent inhibitors of thyroid hormone conjugation *in vitro*, with major concerns regarding the sulfation pathway of iodothyronine.[74,75]

In addition to hydroxy-PCBs, methylsulfonyl-PCBs have been shown to contribute to the effect of PCBs on thyroid hormones. Kato *et al.* tested several methylsulfonyl metabolites of tetra- and pentachlorobiphenyls in rats and found that 3- and 4-methylsulfonyl metabolites of tetra- and pentachlorobiphenyl reduce thyroid hormone levels, suggesting that these metabolites can act as endocrine disruptors.[76] The thyroid is not the only target of methysulfonyl metabolites. Letcher *et al.* found that they exhibit anti-estrogenic activity by competing with estradiol for binding to ERα.[77] Furthermore, methylsulfonyl-PCBs may affect the glucocorticoid system by competitively inhibiting CYP-corticosterone synthesis in mouse adrenocortical cells, while the parent PCBs does not.[78] They broadly impact adrenocortical steroidogenesis, as demonstrated by Xu *et al.*, using human adrenocortical carcinoma cells.[79] They also compete with dexamethasone for binding to the glucocorticoid receptor and act as antagonists of this receptor.[80] In addition of being biologically active, it is noteworthy that methylsulfonyl-PCBs, which are more hydrophobic and resistant to further degradation than hydroxylated PCBs, accumulate in different tissues such as the liver and lung.[81]

All together, these data indicate that PCB metabolites can induce endocrine disruption by a variety of mechanisms, including the alteration of hormone production or metabolism, and direct agonism or antagonism at the level of some steroid receptors. The activity of individual metabolites varies according to the congeners from which they have been produced, but also depends on the stage of development of the target organism, and on the species. The situation is even more complex if we consider that several PCBs are axially chiral molecules. Individual PCB enantiomers interact enantioselectively with macromolecules such as cytochrome P450 enzymes.[82] A few studies have investigated the enantioselective formation of methylsulfonyl and hydroxyl metabolites of PCBs. Not only was a high specificity in the metabolism of enantiomers observed, but a highly enantiospecific liver retention of some methylsulfonyl-PCBs was also reported.[83] Owing to the barrier to rotation around the central C–C bond for some tri- and tetra-*ortho*-substituted PCBs, 19 PCBs exist as stable atropoisomers, seven of which occur in environmental samples;[82] however, there is still a paucity of data relating to bioactivation/detoxication of PCB atropoisomers.

6.4   Polybromodiphenyl Ethers

Polybromodiphenyl ethers (PBDEs) are widely used as additive flame retardants in the manufacture of various plastics, foams, textiles and electronic casings. They show a striking structural resemblance to PCBs and, like their chlorinated homologues, they have been identified as endocrine disruptors.[84] Hormone changes and interactions with nuclear receptors have been reported with PBDE mixtures as well as with some individual congeners.[85] For instance, BDE-19, -49 and -100 are antagonists of the androgen and progesterone receptors, and are ERα agonists. They inhibit estradiol sulfotransferase activity and have potentiating effects on the cell proliferation mediated by T3.[86] Like PCBs, PBDEs can be transformed into hydroxylated metabolites.[87] In rodents, BDE-47 can be transformed into different tri- or tetrabrominated hydroxylated metabolites.[88,89] BDE-99 and -100 have also been found to be metabolized to a number of different OH-PBDEs in rat.[90,91]

Some studies have investigated the endocrine disrupting activity of hydroxylated PBDE metabolites, compared to unchanged PBDEs. The estrogenic and anti-estrogenic potencies of 17 PBDE congeners, as well three hydroxylated PBDEs, have been studied in different cell line assays based on ER-dependent luciferase reporter gene expression, including ER-CALUX assay.[92] In the T47D-Luc-based ER-CALUX assay, 11 PBDEs exhibited luciferase induction in a dose-dependent manner. The most potent congeners (BDE-30, -51, -75, -100 and -119) showed EC_{50} values within a small concentration range of 2.5–3.9 μM. Among the hydroxylated PBDEs tested, 4′-hydroxy-BDE-30, a potential metabolite of BDE-30, showed the highest estrogenic potency (EC_{50} of 0.1 μM). In addition, the agonistic activity of both BDE-30 and the corresponding hydroxylated metabolite were much higher in the ERα- than in the ERβ-luc cell lines. Although based on a limited number of compounds, this work suggests that hydroxylated PBDEs are more active inducers of the ER signal transduction pathway *in vitro* than the corresponding parent PBDEs. In addition, whereas BDE-47 was determined to be a weak ER agonist in the ER-CALUX assay, 6-hydroxy-BDE-47 was found to be an ER antagonist, showing that in this case the biological properties of the metabolite were opposite to that of parent compound, based only on slight structural modifications. Interestingly, 6-hydroxy-BDE-47 was also found to exhibit other endocrine disruptor properties, not involving NR interaction. Indeed, this metabolite was shown to have a much higher affinity than the natural ligand T4 towards the specific hormone transport protein transthyretin, therefore being a potential T4 competitor.[93] The same authors tested the endocrine disrupting potency of several BDE-47 metabolites produced *in vitro* using hepatic microsomes from rats induced by phenobarbital treatment, and compared them to the parent compound. They found that individual metabolites had transthyretin-binding and estradiol sulfotransferase-inhibiting potencies 160–1600 and 2.2–220 times higher than BDE-47 itself, respectively. These recent data confirm that hydroxylation of PBDEs may result in the formation of active metabolites, as previously demonstrated for PCBs.

He *et al.* tested the steroidogenic effects of 20 hydroxylated or methoxylated PBDEs, including 5- and 6-hydroxy-BDE-47 and 5- and 6-methoxy-BDE-47.[94] Several metabolites were found to affect steroidogenesis at the gene, enzyme and hormone levels. CYP11B2, which regulates the synthesis of aldosterone, was the most sensitive gene tested and was induced by most of the metabolites. Some metabolites affected aromatase activity or interfered with testosterone or estradiol production, suggesting that PBDE metabolites may disrupt endocrine function *in vivo*. However, it cannot be concluded whether the biotransformation pathways involved in the production of these metabolites can be considered as a bioactivation process, since the parent compounds have not been tested in parallel.

6.5   Vinclozolin

Vinclozolin is a systemic dicarboximide fungicide used in plant and fruit treatments. The major metabolic pathways of vinclozolin in rodents are summarized in Figure 6.3. The first step is the hydrolytic opening of the oxazolidine

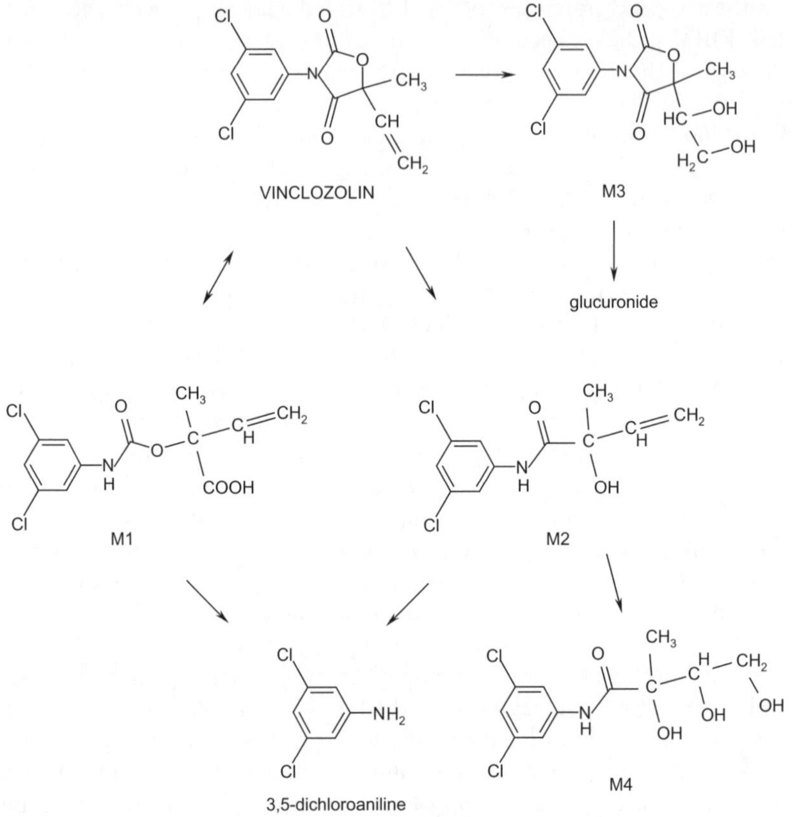

Figure 6.3 Metabolic pathways of vinclozolin in mammals.

ring, involving non-enzymatic cleavage of the C2–N3 or N3–C4 bonds, to give compounds M2 and M1, respectively. Cleavage of the N2–C3 bond is followed by decarboxylation, and is thus not reversible, in contrast to the cleavage of the N3–C4 bond.[95] The enzymatic phase I step includes the dihydroxylation of the vinyl group, presumably *via* an epoxide intermediate, resulting in the bio-transformation of metabolite M2 into M4. The same process may occur directly from vinclozolin (minor pathway not shown in the figure). Phase II pathways include extensive conjugation to glucuronic acid of the dihy-droxylated M3 metabolite. 3,5-Dichloroaniline is also found in minor amounts in metabolic studies performed *in vitro* with vinclozolin.[95] Vinclozolin has been characterized as a potent *in vitro* and *in vivo* androgen antagonist,[96–99] and the active anti-androgenic compounds were identified as metabolites M1 and M2.[100] When administered during sexual differentiation, this fungicide has been shown to demasculinize and feminize male rat offspring.[101,102] Anway and co-workers demonstrated that embryonic exposure to vinclozolin at the time of gonadal sex determination caused a transgenerational effect on male rat fertility and testis function.[103,104]

More recently, vinclozolin and metabolites M1 and M2 have been tested using bioluminescent reporter cell lines, to investigate their capability to activated various nuclear receptors, including androgen (AR), progesterone (PR), glucocorticoid (GR), mineralocorticoid (MR) and ER.[105] As expected, vinclozolin metabolites were found to interact with AR. Transcriptional activity of AR in response to vinclozolin and its metabolites showed that M2 had partial agonistic activity (20% of maximal activity) while vinclozolin and M1 did not. Despite its partial agonistic activity, M2 was a better antagonist than vinclozolin and M1. M2 was also a PR, GR and MR antagonist (MR > > PR > GR). It was also observed that vinclozolin, M1 and M2 were agonists for both ERα and ERβ, with a lower affinity for ERβ. Although these effects have to be confirmed *in vivo*, these data suggest M1 and M2 are able to act through more than one mechanism. The pathways leading to M1 and M2 are clearly bioactivation pathways, since their efficiency in producing an endocrine disruption is equivalent to, or higher than, vinclozolin itself. It has been reported that M3 was a poor inhibitor of androgen receptor binding; in addition, it is rapidly conjugated to glucuronic acid, suggesting that additional metabolism of M2 leads to products that are not anti-androgenic.[95,106] In contrast, 3,5-dichloroaniline is known to be potentially biotransformed by CYP enzymes into an hydroxylamine and subsequently to a genotoxic nitroso intermediate.[107]

6.6   Methoxychlor

Methoxychlor, a structural analog of DDT, has been shown to be responsible for reproductive as well as developmental toxicity in mammals.[108–111] Demethylation is the main biotransformation pathway of methoxychlor, yielding (mono-hydroxy)methoxychlor (mono-OH-Mx) and subsequently (bishydroxy)methoxychlor (bis-OH-Mx; Figure 6.4). CYP2C19, CYP2C9, CYP2C11 and CYP1A2 are the major catalysts of methoxychlor demethylation in humans.[112–114] Bis-OH-Mx can be further metabolized to a trihydroxycatechol metabolite (tris-OH-Mx) or conjugated to glucuronic acid.[115–117] In addition to mono-OH-Mx and bis-OH-Mx, incubation of methoxychlor with liver microsomes from phenobarbital-treated rats yields ring-hydroxylated methoxychlor.[115] These data indicate that there are two possible pathways for the formation of tris-OH-Mx: one by ring hydroxylation of methoxychlor followed by demethylation and the other by demethylation to mono-OH-Mx and bis-OH-Mx, followed by ring hydroxylation (Figure 6.4). The *ortho*-hydroxylation is primarily due to CYP3A4 and CYP2B,[115,118] while CYP2C19, CYP2C11, CYP2C9 and CYP1A2 are predominantly involved in sequential *O*-demethylation.[112,114] *In vivo* studies on methoxychlor metabolism conducted in goats showed that most urinary and fecal methoxychlor metabolites were demethylated and dechlorinated derivatives, a portion of these being conjugated with glucuronic acid.[119] Dehydrochlorinated metabolites were also observed by these authors (not shown in Figure 6.4).

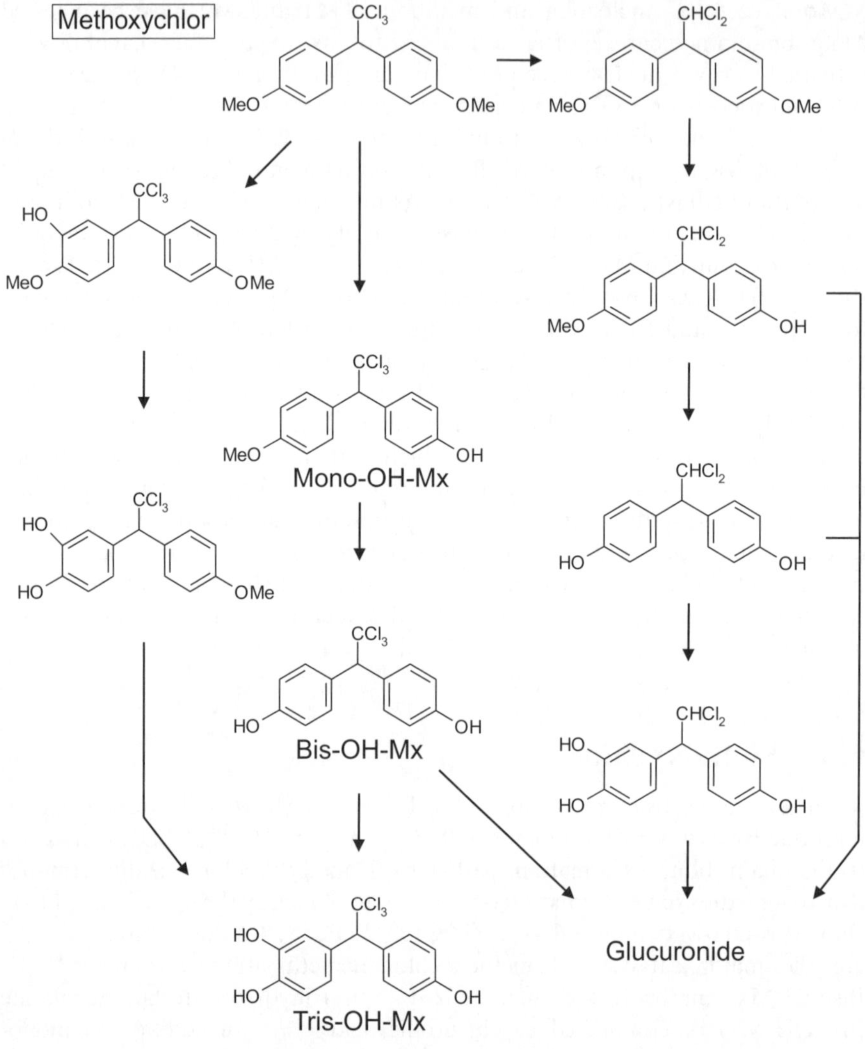

Figure 6.4 *In vitro/in vivo* metabolic pathways of methoxychlor in mammals.

Conversion of methoxychlor to bis-OH-Mx is considered to be a bioactivation pathway since this metabolite is approximately 100 and 10 times more active than the parent compound for ERα and AR, respectively.[120,121] A thorough study, comparing the agonist and antagonist activities of methoxychlor and its major metabolites for ERα, ERβ and AR, demonstrated that bis-OH-Mx, and at a lesser extent mono-OH-Mx, were agonists of ERα but antagonists of both ERβ and AR. Conversely, methoxychlor was found to be a partial agonist for AR.[122] The trihydroxy metabolite of methoxychlor displayed only weak ERα agonist activity. These data confirm that demethylation can be considered as a bioactivation pathway for methoxychlor. In addition, when methoxychlor is administered to rats, the CYP2C subfamily is induced, suggesting that methoxychlor accelerates its own conversion into hormonally active metabolites.[113] The bioactivation of methoxychlor is even more complex. Indeed, mono-OH-Mx is a chiral compound and the demethylation of methoxychlor is enantioselective. Human liver microsomes produce mainly the (*S*)-mono-OH-Mx, and this reaction is primarily catalyzed by CYP2C9.[114] In contrast, recombinant CYP1A2 primarily demethylates methoxychlor into (*R*)-mono-OH-Mx, suggesting that this isoform contributes to a limited extent to the metabolism of methoxychlor in humans. Taking into account the fact that (*S*)-mono-OH-Mx is three times more active than the (*R*)-enantiomer for ERα, as demonstrated by Miyashita and co-workers,[123] the estrogenic activity of methoxychlor after metabolic activation should be higher than estimated from the activity of racemic mixtures.

In addition to its capability to produce various ER- and AR-dependent effects, bis-OH-Mx also inhibits steroidogenesis in rat ovarian granulosa cells,[124] suggesting that the detrimental effects of methoxychlor on female fertility may be primarily due to its metabolites.[125] Recently, Harvey *et al.*, using a genomic approach, demonstrated that bis-OH-Mx exerts multiple effects within the granulosa cell steroidogenic pathway which are not linked to an ER- or AR-mediated mechanism.[126] Methoxychlor and its metabolite bis-OH-Mx can also alter normal embryonic testis development and growth.[127] Bis-OH-Mx has been shown to inhibit testosterone formation in a dose-dependent manner. More specifically, the major effect of this metabolite was found to be an inhibition of the metabolic step which consists in cholesterol conversion into pregnenolone, through the cleavage of the side-chain of cholesterol.[128] This effect did not appear to be mediated through ER or AR.

The example of methoxychlor demonstrates that biotransformation can influence the endocrine disrupting properties of chemicals through multiple pathways. These include the formation of metabolites with increased potency and efficacy, the production of metabolites which can act on molecular targets that are distinct from that of the parent molecule, or of metabolites which exhibit an increased potency at the level of one receptor, while being antagonists at the level of other receptors.[129] The metabolism of methoxychlor also provides evidence that both the enantioselectivity of the reaction and the specific activation of ER by each enantiomer are parameters that should be taken into account for some xenobiotics.

6.7 Conclusions

In this chapter we have focused on the bioactivation processes of selected xenobiotics, which are among the best known examples of metabolic activation, or are molecules for which the mechanisms of bioactivation have been subject to detailed *in vitro/in vivo* studies. Other food contaminants/components are known to be activated into endocrine disrupting compounds, through metabolic processes. These includes pesticides (*e.g.* biphenyl, DDT, fipronil), polycyclic aromatic hydrocarbons (benzo[*a*]pyrene, *etc.*), nitropolycyclic aromatic hydrocarbons (nitrofluorene, *etc.*), plasticizers (phthalates, *etc.*) or naturally occurring compounds such as phytoestrogens or mycotoxins (*e.g.* chalcones, zearalenone). As illustrated by the examples developed in this chapter, in most cases the bioactivation process involves cytochrome P450. However, some exceptions exist, as demonstrated, for instance, by the formation of equol from daidzein, which implies the participation of reductases expressed by the gut microflora. Another well-known example is the formation of monoethylhexyl phthalate (which is the active compound produced from diethylhexyl phthalate), and which is catalyzed by esterases. Phase II enzymes are generally regarded as detoxification reactions, although some metabolites originating from glutathione conjugation and subsequent mercapturic acid formation, such as methylsulfonyl derivatives of PCBs, are active on different hormonal targets.

Because some endocrine disrupting compounds are active only after biotransformation, it is important to know the metabolic competence of the models used to test their activity. Many of the cell lines used for such purposes express only part (or none) of the metabolic activities expressed in the tissue they originate from. The limited capabilities of *in vitro* models to metabolically activate or inactivate xenobiotics may lead to a misinterpretation of *in vitro* results, if such information is not taken into account.[130] In addition, the different possible modes of actions of biotransformation products (*i.e.* activation of nuclear receptors, synthesis or degradation of hormones, binding to transport proteins) implies the need to use a series of appropriate complementary assays before a sound assessment of the potential endocrine disrupting effects of a given compound is possible. Moreover, the fact that certain enzymes can specifically catalyze the formation of selected enantiomers, and given interaction with the receptor can also be enantiomer specific (as shown for methoxychlor metabolites), suggest that particular attention should be paid to the analysis of compounds with chiral centers.

References

1. D. W. Nebert and D. W. Russell, *Lancet*, 2002, **360**, 1155.
2. F. P. Guengerich, *J. Biochem. Mol. Toxicol.*, 2007, **21**, 163.
3. F. P. Guengerich, *Chem. Res. Toxicol.*, 2008, **21**, 70.
4. U. A. Boelsterli, *Mechanistic Toxicology: The Molecule Basis of How Chemicals Disrupt Biological Targets*, Taylor & Francis, London, 2003.
5. F. P. Guengerich, in *Cytochrome P450: Structure, Mechanism, and Biochemistry*, ed. P. R. Ortiz de Montellano, Kluwer/Plenum, New York, 2005, p. 377.
6. A. Parkinson and B. W. Ogilvie, in *Casarett and Doull's Toxicology: The Basic Science of Poisons*, ed. L. J. Casarett, D. Curtis, C. D. Klaassen and J. Doull, McGraw-Hill, New York, 2008, p. 161.
7. C. D. Klaassen, *Toxicol. Appl. Pharmacol.*, 1973, **24**, 37.

8. D. W. Nebert and V. Vasiliou, *Hum. Genomics*, 2004, **1**, 460.
9. W. Dekant, in *Molecular, Clinical and Environmental Toxicology*, ed. A. Luch, Birkhäuser, Basel, 2009, p. 57.
10. B. Sallustio, in *Advances in Molecular Toxicology*, ed. J. C. Fishbein, Elsevier, Amsterdam, 2008, vol. 2, p. 57.
11. P. J. Van Bladeren, *Chem. Biol. Interact.*, 2000, **129**, 61.
12. J. E. Bakke and J. A. Gustafsson, *Xenobiotica*, 1986, **16**, 1047.
13. A. Hillenweck, J. P. Cravedi, L. Debrauwer, J. C. Killeen Jr., M. Bliss Jr. and D. E. Corpet, *Pest. Biochem. Physiol.*, 1997, **58**, 34.
14. S. Kitamura, K. Sugihara, S. Sanoh, N. Fujimoto and S. Ohta, *J. Health. Sci.*, 2008, **54**, 343.
15. E. Diamanti-Kandarakis, J. P. Bourguignon, L. C. Giudice, R. Hauser, G. S. Prins, A. M. Soto, R. T. Zoeller and A. C. Gore, *Endocrin. Rev.*, 2009, **30**, 293.
16. K. D. R. Setchell, N. M. Brown and E. Lydeking-Olsen, *J. Nutr.*, 2002, **132**, 3577.
17. J. P. Antignac, R. Cariou, B. Le Bizec, J. P. Cravedi and F. André, *Rapid Commun. Mass Spectrom.*, 2003, **17**, 1256.
18. A. Hoikkala, E. Mustonen, I. Saastomoinen, T. Jokela, J. Taponen, H. Saloniemi and K. Wähälä, *Mol. Nutr. Food Res.*, 2007, **51**, 782.
19. K. D. R. Setchell and C. Clerici, *J. Nutr.*, 2010, **140**, 1355S.
20. R. A. King and D. B. Bursill, *Am. J. Clin. Nutr.*, 1998, **67**, 867.
21. S. R. Shelnutt, C. O. Cimino, P. A. Wiggins and T. M. Badger, *Cancer Epidemiol. Biomarkers Prev.*, 2000, **9**, 413.
22. K. D. R. Setchell, N. M. Brown, P. Desai, L. Zimmer-Nechemias, B. E. Wolfe, W. T. Brashear, A. S. Kirschner, A. Cassidy and J. E. Heubi, *J. Nutr.*, 2001, **131**, S1362.
23. S. Vergne, K. Titier, V. Bernard, J. Asselineau, M. Durand, V. Lamothe, M. Potier, P. Perez, J. Demotes-Mainard, P. Chantre, N. Moore, C. Bennetau-Pelissero and P. Sauvant, *J. Pharm. Biomed. Anal.*, 2007, **43**, 1488.
24. C. Clerici, K. D. R. Setchell, P. M. Battezzati, M. Pirro, V. Giuliano, S. Asciutti, D. Castellani, E. Nardi, G. Sabatino, S. Orlandi, M. Baldoni, O. Morelli, E. Mannarino and A. Morelli, *J. Nutr.*, 2007, **137**, 2270.
25. K. D. R. Setchell, N. M. Brown, L. Zimmer-Nechemias, W. T. Brashear, B. Wolfe, A. S. Kirschner and J. E. Heubi, *Am. J. Clin. Nutr.*, 2002, **76**, 447.
26. A. J. Day, M. S. Dupont, S. Ridley, M. Rhodes, M. J. Rhodes, M. R. Morgan and G. Willamson, *FEBS Lett.*, 1998, **436**, 71.
27. S. E. Kulling, L. Lehmann and M. Metzler, *J. Chromatogr. B*, 2002, **777**, 211.
28. S. M. Heinonen, K. Wähälä and H. Adlercreutz, *J. Agric. Food Chem.*, 2004, **52**, 6802.
29. M. S. Kurzer and X. Xu, *Annu. Rev. Nutr.*, 1997, **17**, 353.
30. C. E. Rüfer, H. Glatt and S. E. Kulling, *Drug Metab. Dispos.*, 2006, **34**, 51.

31. D. R. Doerge, H. C. Chang, M. I. Churchwell and C. L. Holder, *Drug Metab. Dispos.*, 2000, **28**, 298.
32. V. Breinholt and J. C. Larsen, *Chem. Res. Toxicol.*, 1998, **1**, 622.
33. V. Breinholt, A. Hossaini, G. W. Svendsen, C. Brouwer and S. E. Nielsen, *Food Chem. Toxicol.*, 2000, **38**, 555.
34. C. Atkinson, C. L. Frankenfeld and J. W. Lampe, *Exp. Biol. Med.*, 2005, **230**, 155.
35. K. Morito, T. Hirose, T. Kinjo, T. Hirakawa, M. Okawa, T. Nohara, S. Ogawa, S. Inoue, M. Muramatsu and Y. Masamune, *Biol. Pharm. Bull.*, 2001, **24**, 351.
36. D. Kostelac, G. Rechkemmer and K. Brivida, *J. Agric. Food Chem.*, 2003, **51**, 7632.
37. R. S. Muthyala, H. Y. Ju, S. Sheng, L. D. Williams, D. R. Doerge, B. S. Katzenellenbogen, W. G. Helferich and J. A. Katzenellenbogen, *Bioorg. Med. Chem.*, 2004, **12**, 1559.
38. J. Kinjo, R. Tsuchihashi, K. Morito, T. Hirose, T. Aomori, T. Nagao, H. Okabe, T. Nohara and Y. Masamune, *Biol. Pharm. Bull.*, 2004, **27**, 185.
39. N. Sathyamoorthy and T. T. Wang, *Eur. J. Cancer*, 1997, **33**, 2384.
40. E. Schmitt, W. Dekant and H. Stopper, *Toxicol. In Vitro*, 2001, **15**, 433.
41. M. E. Martin, M. Haourigui, C. Pelissero, C. Benassayag and E. A. Nunez, *Life Sci.*, 1996, **58**, 429.
42. T. D. Lund, D. J. Munson, M. E. Haldy, K. D. Setchell, E. D. Lephart and R. J. Handa, *Biol. Reprod.*, 2004, **70**, 1188.
43. S. O. Mueller, S. Simon, K. Chae, M. Metzler and K. S. Korach, *Toxicol. Sci.*, 2004, **80**, 14.
44. I. C. Munro, M. Harwood, J. J. Hlywka, A. M. Stephen, J. Doull, W. G. Flamm and H. Adlercreutz, *Nutr. Rev.*, 2003, **61**, 1.
45. L. Lehmann, H. L. Esch, J. Wagner, L. Rohnstock and M. Metzler, *Toxicol. Lett.*, 2005, **158**, 72.
46. M. Bilau, I. Sioen, C. Matthys, A. De Vocht, G. Goemans, C. Belpaire, J. L. Willems and S. De Henauw, *Food Addit. Contam.*, 2007, **24**, 1386.
47. E. Fattore, R. Fanelli, E. Dellatte, A. Turrini and A. di Domenico, *Chemosphere*, 2008, **73**, S278.
48. V. Sirot, A. Tard, P. Marchand, B. Le Bizec, A. Venisseau, A. Brosseau, J. L. Volatier and J. C. Leblanc, *Organohalogen Compd.*, 2006, **68**, 383.
49. EFSA, *EFSA J.*, 2005, 284, 1; http://www.efsa.eu.int.
50. I. A. T. M. Meerts, Y. Assink, P. H. Cenijn, J. H. Van Den Berg, B. M. Weijers, A. Bergman, J. H. Koeman and A Brouwer, *Toxicol. Sci.*, 2002, **68**, 361.
51. D. Meironyté Guvenius, A. Aronsson, G. Ekman-Ordeberg, A. Bergman and K. Norén, *Environ. Health Perspect.*, 2003, **111**, 1235.
52. S. D. Soechitram, M. Athanasiadou, L. Hovander, Å. Bergman and P. J. J. Sauer, *Environ. Health Perspect.*, 2004, **112**, 1208.
53. M. O. James, in *PCBs: Recent Advances in Environmental Toxicology and Health Effects*, ed., L.W. Robertson and L. G. Hansen, University Press of Kentucky, Lexington, 2001, p. 35.

54. G. G. Oakley, L. W. Robertson and R. C. Gupta, *Carcinogenesis*, 1996, **17**, 109.
55. K. S. Korach, P. Sarver, K. Chae, J. A. McLachlan and J. D. McKinney, *Mol. Pharmacol.*, 1988, **33**, 120.
56. S. Safe, in *The Handbook of Environmental Chemistry, Vol. 3: Endocrine Disruptors, Part I*, ed. M. Metzler, Springer, Berlin, 2001, p. 155.
57. M. Moore, M. Mustain, K. Daniel, I. Chen, S. Safe, T. Zacharewsky, B. Gillesby, A. Joyeux and P. Balaguer, *Toxicol. Appl. Pharmacol.*, 1997, **142**, 160.
58. M. Pliskova, J. Vondracek, R. Fernandez Canton, J. Nera, A. Kocan, J. Petrik, T. Trnovec, T. Sanderson, M. van den Berg and M. Machala, *Environ. Health Perspect.*, 2005, **113**, 1277.
59. G. G. J. M. Kuiper, J. G. Lemmen, B. Carlsson, J. C. Corton, S. H. Safe, P. T. van der Saag, B. van der Burg and J. A. Gustafsson, *Endocrinology*, 1998, **139**, 4252.
60. M. R. Fielden, I. Chen, B. Chittim, S. H. Safe and T. R. Zacharewski, *Environ. Health Perspect.*, 1997, **105**, 1238.
61. K. Connor, K. Ramamoorthy, M. Moore, M. Mustain, I. Chen, S. Safe, T. Zacharewski, B. Gillesby, A. Joyeux and P. Balaguer, *Toxicol. Appl. Pharmacol.*, 1997, **145**, 111.
62. M. H. A. Kester, S. Bulduk, D. Tibboel, W. Meinl, H. Glatt, C. N. Falany, M. A. H. Coughtrie, A. Bergman, S. Safe, G. G. Kuiper, A. G. Schuur, A. Brouwer and T. J. Visser, *Endocrinology*, 2000, **141**, 1897.
63. M. H. A. Kester, S. Bulduk, H. van Toor, D. Tibboel, W. Meinl, H. Glatt, C. N. Falany, M. A. H. Coughtrie, A. G. Schuur, A. Brouwer and T. J. Visser, *J. Clin. Endocrinol. Metab.*, 2002, **87**, 1142.
64. Y. Liu, J. T. Smart, Y. Song, H. J. Lehmler, L. W. Robertson and M. W. Duffel, *Drug Metab. Dispos.*, 2009, **37**, 1065.
65. P. S. Cooke, T. Sato and D. L. Buchanan, in *PCBs: Recent Advances in Environmental Toxicology and Health Effects*, ed. L.W. Robertson and L. G. Hansen, University Press of Kentucky, Lexington, 2001, p. 257.
66. C. E. Garner, W. N. Jefferson, L. T. Burka, H. B. Matthews and R. R. Newbold, *Toxicol. Appl. Pharmacol.*, 1999, **154**, 188.
67. C. E. Garner, L. T. Burka, A. E. Etheridge and H. B. Matthews, *Toxicol. Appl. Pharmacol.*, 2000, **162**, 115.
68. P. W. Ho, C. E. Garner, J. W. Ho, K. C. Leung, A. C. Chu, K. H. Kwok, M. H. Kung, L. T. Burka, D. B. Ramsden and S. L. Ho, *Curr. Drug Metab.*, 2008, **9**, 304.
69. R. Kamata, F. Shiraishi, D. Nakajima, H. Takigami and H. Shiraishi, *Toxicol. In Vitro*, 2009, **23**, 736.
70. R. T. Zoeller, in *PCBs: Recent Advances in Environmental Toxicology and Health Effects*, ed. L. W. Robertson and L. G. Hansen, University Press of Kentucky, Lexington, 2001, pp. 265–271.
71. Y. Kato, K. Haraguchi, M. Kubota, Y. Seto, S. Ikushiro, T. Sakaki, N. Koga, S. Yamada and M. Degawa, *Drug Metab. Dispos.*, 2009, **37**, 2095.

72. T. Otake, J. Yoshinaga, T. Enomoto, M. Matsuda, T. Wakimoto, M. Ikegami, E. Suzuki, H. Naruse, T. Yamanaka, N. Shibuya, T. Yasumizu and N. Kato, *Environ. Res.*, 2007, **105**, 240.

73. A. O. Cheek, K. Kow, J. Chen and J. A. McLachlan, *Environ. Health Perspect.*, 1999, **107**, 273.

74. A. G. Schuur, F. F. Legger, M. E. van Meeteren, M. J. H. Moonen, I. van Leeuwen-Bol, A. Bergman, T. J. Visser and A. Brouwer, *Chem. Res. Toxicol.*, 1998, **11**, 1075.

75. A. G. Schuur, A. Bergman, A. Brouwer and T. J. Visser, *Toxicol. In Vitro*, 1999, **13**, 417.

76. Y. Kato, K. Haraguchi, T. Shibahara, S. Yumoto, Y. Masuda and R. Kimura, *Toxicol. Sci.*, 1999, **48**, 51.

77. R. J. Letcher, J. G. Lemmen, B. van der Burg, A. Brouwer, A. Bergman, J. P. Giesy and M. van den Berg, *Toxicol. Sci.*, 2002, **69**, 362.

78. M. Johansson, C. Larsson, A. Bergman and B. O. Lund, *Pharmacol. Toxicol.*, 1998, **83**, 225.

79. Y. Xu, R. M. K. Yu, X. Zhang, M. B. Murphy, J. P. Giesy, M. H. W. Lam, P. K. S. Lam, R. S. S. Wu and H. Yu, *Chemosphere*, 2006, **63**, 772.

80. M. Johansson, S. Nilsson and B. O. Lund, *Environ. Health Perspect.*, 1998, **106**, 769.

81. R. J. Letcher, E. Klasson-Wehler and A. Bergman, in *The Handbook of Environmental Chemistry: New Types of Persistent Halogenated Compounds*, ed. J. Paasivirta, Springer, Berlin, 2000, vol. 3, p. 315.

82. H. J. Lehmler and L. W. Robertson, in *PCBs: Recent Advances in Environmental Toxicology and Health Effects*, ed. L.W. Robertson and L. G. Hansen, University Press of Kentucky, Lexington, 2001, p. 61.

83. H. J. Lehmler, S. J. Harrad, H. Hühnerfuss, I. Kania-Korwel, C. M. Lee, Z. Lu and C. S. Wong, *Environ. Sci. Technol.*, 2010, **44**, 2757.

84. J. Legler and A. Brouwer, *Environ. Int.*, 2003, **29**, 879.

85. T. E. Stoker, R. L. Cooper, C. S. Lambright, V. S. Wilson, J. Furr and L. E. Gray, *Toxicol. Appl. Pharmacol.*, 2005, **207**, 78.

86. T. Hamers, J. H. Kamstra, E. Sonneveld, A. J. Murk, M. H. A. Kester, P. L. Andersson, J. Legler and A. Brouwer, *Toxicol. Sci.*, 2006, **92**, 157.

87. H. Hakk and R. J. Letcher, *Environ. Int.*, 2003, **29**, 801.

88. U. Örn and E. Klasson-Wehler, *Xenobiotica*, 1998, **28**, 199.

89. G. Marsh, M. Athanasiadou, I. Athanasiadis and A. Sandholm, *Chemosphere*, 2006, **63**, 690.

90. H. Hakk, G. Larsen and E. Klasson-Wehler, *Xenobiotica*, 2002, **32**, 369.

91. H. Hakk, J. Huwe, M. Low, D. Rutherford and G. Larsen, *Xenobiotica*, 2006, **36**, 79.

92. I. A. T. M. Meerts, R. J. Letcher, S. Hoving, G. Marsh, A. Bergman, J. G. Lemmen, B. Van den Burg and A. Brouwer, *Environ. Health Perspect.*, 2001, **109**, 399.

93. T. Hamers, J. H. Kamstra, E. Sonneveld, A. J. Murk, T. J. Visser, M. J. M. Van Velzen, A. Brouwer and A. Bergman, *Mol. Nutr. Food Res.*, 2008, **52**, 284.

94. Y. He, M. B. Murphy, R. M. K. Yu, M. H. W. Lam, M. Hecker, J. P. Giesy, R. S. S. Wu and P. K. S. Lam, *Toxicol. Lett.*, 2008, **176**, 230.

95. J. Bursztyka, L. Debrauwer, E. Perdu, I. Jouanin, J. P. Jaeg and J. P. Cravedi, *J. Agric. Food Chem.*, 2008, **56**, 4832.

96. W. R. Kelce, C. R. Lambright, L. E. Gray Jr. and K. P. Roberts, *Toxicol. Appl. Pharmacol.*, 1997, **142**, 192.

97. L. E. Gray Jr., J. Ostby, E. Monosson and W. R. Kelce, *Toxicol. Ind. Health*, 1999, **15**, 48.

98. E. Monosson, W. R. Kelce, C. Lambright, J. Ostby and L. E. Gray Jr., *Toxicol. Ind. Health*, 1999, **15**, 65.

99. C. Nellemann, M. Dalgaard, H. R. Lam and A. M. Vinggaard, *Toxicol. Sci.*, 2003, **71**, 251.

100. C. Wong, W. R. Kelce, M. Sar and E. M. Wilson, *J. Biol. Chem.*, 1995, **270**, 19998.

101. L. E. Gray Jr., J. Ostby and W. R. Kelce, *Toxicol. Appl. Pharmacol.*, 1994, **129**, 46.

102. T. Shono, S. Suita, H. Kai and Y. Yamaguchi, *J. Pediatr. Surg.*, 2004, **39**, 213.

103. M. D. Anway, A. S. Cupp, M. Uzumcu and M. K. Skinner, *Science*, 2005, **308**, 1466.

104. M. D. Anway, C. Leathers and M. K. Skinner, *Endocrinology*, 2006, **147**, 5515.

105. J. M. Molina-Molina, A. Hillenweck, I. Jouanin, D. Zalko, J. P. Cravedi, M. F. Fernandez, A. Pillon, J. C. Nicolas, N. Olea and P. Balaguer, *Toxicol. Appl. Pharmacol.*, 2006, **216**, 44.

106. W. R. Kelce, E. Monosson, M. P. Gamcsik, S. C. Laws and L. E. Gray Jr., *Toxicol. Appl. Pharmacol.*, 1994, **126**, 276.

107. EPA, Vinclozolin, 2000; http://www.epa.gov/oppsrrd1/REDs/2740red.pdf.

108. R. E. Chapin, M. W. Harris, B. J. Davis, S. M. Ward, R. E. Wilson, M. A. Mauney, A. C. Lockhart, R. J. Smialowicz, W. C. Moser, L. T. Burka, B. J. Collins, E. A. Haskins, J. D. Allen, L. Judd, W. A. Purdie, H. L. Harris, C. A. Lee and G. M. Corniffe, *Fundam. Appl. Toxicol.*, 1997, **40**, 138.

109. P. S. Cooke and V. P. Eroschenko, *Biol. Reprod.*, 1990, **42**, 585.

110. D. L. Hall, L. A. Payne, J. M. Putnam and Y. M. Huet-Hudson, *Reprod. Toxicol.*, 1997, **11**, 703.

111. E. M. Martinez and W. J. Swartx, *Reprod. Toxicol.*, 1991, **5**, 139.

112. D. M. Stresser and D. Kupfer, *Drug Metab. Dispos.*, 1998, **26**, 868.

113. E. Mikamo, S. Harada, J. Nishikawa and T. Nishihara, *J. Health Sci.*, 2003, **49**, 229.

114. Y. Hu, K. Krausz, H. V. Gelboin and D. Kupfer, *Xenobiotica*, 2004, **34**, 117.

115. S. S. Dehal and D. Kupfer, *Drug Metab. Dispos.*, 1994, **22**, 937.

116. E. Hazai, P. V. Gagne and D. Kupfer, *Drug Metab. Dispos.*, 2004, **32**, 742.

117. D. Kupfer, W. H. Bulger and A. D. Theoharides, *Chem. Res. Toxicol.*, 1990, **3**, 8.

118. Y. Hu and D. Kupfer, *Drug Metab. Dispos.*, 2002, **30**, 1329.
119. K. L. Davison, V. J. Feil and C. J. H. Lamoureux, *J. Agric. Food Chem.*, 1982, **30**, 130.
120. K. W. Gaido, L. S. Leonard, S. C. Maness, J. M. Hall, D. P. McDonnell, B. Saville and S. Safe, *Endocrinology*, 1999, **140**, 5746.
121. S. C. Maness, D. P. McDonnell and K. W. Gaido, *Toxicol. Appl. Pharmacol.*, 1998, **151**, 135.
122. K. W. Gaido, S. C. Maness, D. P. McDonell, S. S. Dehal, D. Kupfer and S. Safe, *Mol. Pharmacol.*, 2000, **58**, 852.
123. M. Miyashita, T. Shimada, S. Nakagami, N. Kurihara, H. Miyagawa and M. Akamatsu, *Chemosphere*, 2004, **54**, 1273.
124. R. Zachow and M. Uzumcu, *Reprod. Toxicol.*, 2006, **22**, 659.
125. L. E. Gray Jr., J. Ostby, J. Ferrell, G. Rehnberg, R. Linder, R. Cooper, J. Goldman, V. Slott and J. Laskey, *Fundam. Appl. Toxicol.*, 1989, **12**, 92.
126. C. N. Harvey, M. Esmail, Q. Wang, A. I. Brooks, R. Zachow and M. Uzumcu, *Toxicol. Sci.*, 2009, **110**, 95.
127. A. S. Cupp and M. K. Skinner, *Reprod. Toxicol.*, 2001, **15**, 317.
128. E. P. Murono and R. C. Derk, *Reprod. Toxicol.*, 2005, **20**, 503.
129. M. N. Jacobs, W. Janssens, U. Bernauer, E. Brandon, S. Coecke, R. Combes, P. Edwards, A. Freidig, A. Freyberger, R. Kolanczyk, C. McArdle, O. Mekenyan, P. Schmieder, T. Schrader, M. Takeyoshi and B. van der Burg, *Curr. Drug Metab.*, 2008, **9**, 796.
130. J. Bursztyka, E. Perdu, K. Pettersson, I. Pongratz, M. Fernández-Cabrera, N. Olea, L. Debrauwer, D. Zalko and J. P. Cravedi, *Toxicol. in Vitro*, 2008, **22**, 1595.

Aryl Hydrocarbon Receptor Targeted by Xenobiotic Compounds and Dietary Phytochemicals

JASON MATTHEWS

Department of Pharmacology & Toxicology, University of Toronto, Toronto, Ontario, M5S 1A8, Canada

7.1 Introduction

The aryl hydrocarbon receptor (AhR), also known as the dioxin receptor, is a ligand-activated transcription factor and a member of the basic helix–loop–helix/PER-ARNT-SIM (bHLH-PAS) family, which are often thought of as "biological sensors".[1] AhR is activated or inactivated by a variety of xenobiotic, exogenous and endogenous chemicals; however, the clear candidate endogenous ligand has not been definitively identified.[2–4] AhR mediates the toxic action of many halogenated aromatic hydrocarbons (HAHs), including polychlorinated dibenzo-*p*-dioxins (PCDDs), polychlorinated dibenzofurans (PCDFs) and co-planar polychlorinated biphenyls (PCBs). Detectable levels of these contaminants are present in all humans.[5] These chemicals are highly toxic, persistent and bioaccumulative environmental contaminants, with TCDD being the most toxic compound.[6,7] TCDD has a half-life of approximately 7.5 years in humans and is classified by the International Agency for

Issues in Toxicology No. 11
Hormone-Disruptive Chemical Contaminants in Food
Edited by Ingemar Pongratz and Linda Vikström Bergander

Published by the Royal Society of Chemistry, www.rsc.org

Research on Cancer as a group I carcinogen (known to cause cancer in humans), a classification that was recently reconfirmed.[8] The U.S. Environmental Protection Agency (U.S. EPA) has classified TCDD as carcinogenic to humans and the agency has characterized complex mixtures of dioxins as likely human carcinogens.[9] Laboratory animals exposed to levels of TCDD exhibit a diverse array of species- and tissue-specific toxic and biological effects.[10,11] It is estimated that 90% of human exposure to dioxin-like compounds comes from dietary intake of contaminated food.[12] AhR exhibits profound ligand-binding diversity, binding not only dioxin-like compounds and polycyclic aromatic hydrocarbons (PAHs) but also numerous phytochemicals, such as flavonoids, carotenoids and indoles.[2] Some investigators suggest that 98% of our total dietary intake of AhR-activating compounds comes from phytochemicals.[13] Although some phytochemicals are reported to activate AhR,[14,15] most of them act as antagonists, competing with the classical ligands for binding to AhR and have been suggested to counteract the toxic effects of AhR-activating food contaminants.[16]

7.2 Aryl Hydrocarbon Receptor

AhR is a modular protein with distinct functional domains and the only member of the bHLH-PAS family that exhibits ligand-dependent activation (Figure 7.1).[17] The PAS domain refers to ~275 amino acids shared among multiple family members. Within these 275 amino acid residues are two ~70 amino acid repeats called PAS A and PAS B, which have overlapping but also distinct functions.[1] AhR is a highly conserved protein across vertebrate species and related proteins have been identified in invertebrates, including *Caenorhabditis elegans* and *Drosophila* species.[18] The strong evolutionary conservation of AhR suggests an important biological role;[18,19] however, the physiological role of AhR is unknown. Disruption of AhR gene expression in mice leads to smaller livers, decreased body weight, reduced fecundity and altered female reproduction, indicating an important role for AhR in normal metabolism.[20–22] *Ahr*-null mice are also resistant to many of the

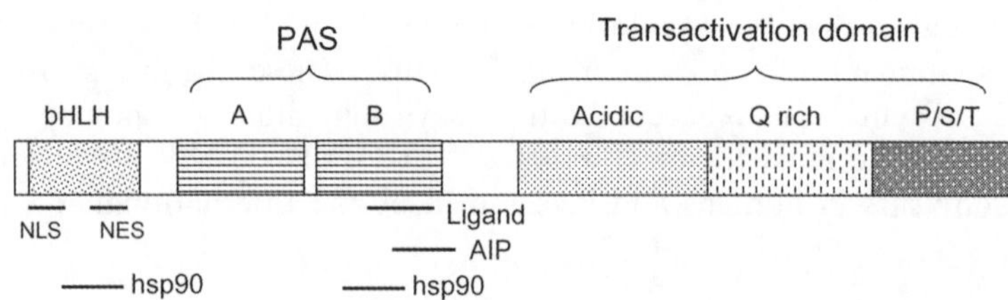

Figure 7.1 Functional domains of the aryl hydrocarbon receptor (AhR). Hsp90, heat shock protein; bHLH, basic helix–loop–helix; NLS, nuclear localization signal; NES, nuclear export signal; Q, glutamine; P/S/T, proline/serine/ threonine domains; AIP, AhR interacting protein.

responses typically observed following treatment with TCDD and related compounds.[23]

The human *AHR* gene is located on chromosome 7p15 and includes 12 exons, 11 of which give rise to a protein with a theoretical mass of ~96 kDa.[24] Genetic variations in AhR among laboratory animals result in altered ligand binding and transcriptional regulation by the receptor, leading to a wide range of sensitivity to the toxic effects of TCDD.[25] One of the well-studied examples of this diversity is the difference in TCDD responsiveness of the C57BL/6J mice relative to the non-responsive DBA/2 mice.[26] The differences in sensitivity are due to an alanine to valine change at amino acid 375 within the ligand binding cavity that results in an approximate 10-fold decrease in binding affinity for TCDD.[27] Altered binding affinity cannot solely explain the vast difference in dioxin sensitivity among or within species. The TCDD-resistant Han/Wister strain of rat has an intronic mutation that causes the deletion of 38 or 43 amino acids from the transactivation domain.[28] The resulting change influences the regulation of key genes involved in TCDD toxicity.[29] However, the TCDD-dependent induction of a prototypical AhR target gene, *CYP1A1*, and affinity for TCDD are virtually identical between the Han/Wistar and the TCDD-sensitive Sprague-Dawley rat strains.[28]

7.2.1 Mechanism of AhR Action

AhR-mediated regulation of CYP1A1 is generally used a model of AhR-dependent transactivation and has been observed in most species.[30] The current, albeit somewhat simplistic, model of AhR action is shown in Figure 7.2. AhR exists in the cytoplasm bound to a chaperone protein complex, containing heat shock protein 90 (Hsp90), AhR interacting protein (AIP, also known as ARA9) and a 23-kDa co-chaperone protein, p23.[31–33] These chaperone proteins help AhR fold and maintain its cytoplasmic location. The PAS B domain partly mediates interactions between AhR and the chaperone proteins. The ligand-binding pocket of AhR is also located within the PAS B domain. After binding ligand, AhR undergoes a conformational change, translocates to the nucleus and binds another bHLH-PAS member and its obligatory dimerization partner, the AhR nuclear translocator (ARNT). This interaction is mediated through the AhR PAS A domain. An N-terminal nuclear export signal shuttles nuclear AhR molecules that fail to interact with ARNT back to the cytoplasm, resulting in its ubiquitination and proteolytic degradation.[34] The AhR/ARNT heterodimer binds to genomic enhancer elements, termed xenobiotic response elements or dioxin response elements (XREs or DREs; 5′-GCGTG-3′) (Figure 7.2).[30] DNA binding results in the recruitment of coregulator proteins (coactivators and corepressors), causing changes in target gene expression.[35] AhR target genes encode proteins that are involved in metabolism, proliferation and differentiation. Well-studied AhR-regulated genes include the drug-metabolizing enzymes cytochrome P4501A1 (CYP1A1), CYP1A2, CYP1B1 and NAD(P):quinone-oxidoreductase (NQO1), which are induced in several

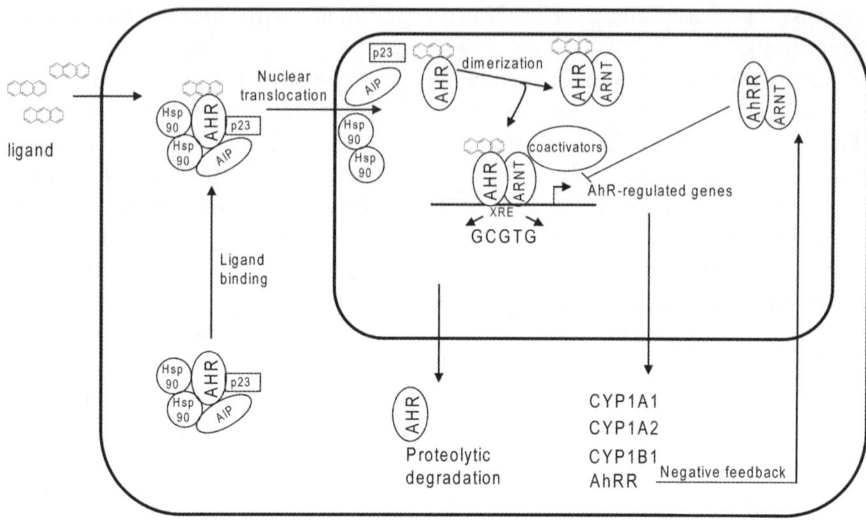

Figure 7.2   Schematic of AhR activation. See text for details.

species after exposure to TCDD and other dioxin-like compounds.[30,36] AhR also regulates the expression of a number of cell cycle genes, including p21, cyclinD1 and cyclinG2.[37–40] The carboxy terminal half of AhR consists of a transactivation region that is divided into three subdomains: (i) the acidic, (ii) glutamine (Q)-rich and (iii) proline (P)-, serine (S)- and threonine (T)-rich domains. This region is the site of interaction with coregulator proteins. As with other transcription factors, ligand-induced recruitment of coregulator proteins is also critical for AhR-mediated transcription and variations in coactivator expression may underlie tissue-specific differences in ligand sensitivity and target gene expression.[35] The recent cloning and characterization of the AhR repressor (AhRR), an AhR-responsive target gene, has led to the suggestion of an autoregulatory negative feedback loop.[41,42] AhRR contains both a bHLH and PAS A domain, but not a PAS B domain. AhR induces expression of AhRR, which then binds and sequesters ARNT, resulting in the quenching of AhR-mediated transcription.[41] Recent studies suggest that AhRR also acts as a transcriptional repressor for other nuclear transcription factors.[43]

In addition to mediating the toxic responses of TCDD and other dioxin-like compounds, AhR regulates the adaptive metabolic response to PAHs, such as benzo[*a*]pyrene. Like TCDD, benzo[*a*]pyrene-activated AhR binds to XREs within the regulatory region of phase I and phase II xenobiotic metabolizing enzymes, including CYP1A1.[36] Benzo[*a*]pyrene and other AhR activating compounds are metabolized by the CYP1 family of enzymes, resulting in their elimination. The ability of ligand-activated AhR to mount an adaptive response to xenobiotic compounds that leads to their degradation has many investigators classifying it as a xenosensor receptor.[44] As a xenosensor, AhR activates a metabolizing enzyme pathway which ultimately leads to the metabolism and elimination of the xenobiotics. The metabolic resilience and stability of TCDD is thought to be the principal determinant of its toxicity due to prolonged and uninterrupted AhR action.[45] Many carcinogens are metabolized by CYP enzymes to either inactivate them or generate chemically reactive electrophiles that covalently bind to DNA. These reactive metabolites may undergo further metabolism by other phase I and II enzymes to inactive compounds.[46] Sustained enzyme activation causes increased metabolism of benzo[a]pyrene and consequently the generation of toxic benzo[a]pyrene metabolites that directly bind to DNA, causing mutations and increasing the risk of cancer. Moreover, CYP1A1-dependent metabolism of benzo[a]pyrene also leads to the generation of reactive oxygen species (ROS).[47] ROS are crucial for life as a number of biochemical reactions require them. All cells are equipped with a multiple-level antioxidant control system that protects the cell from damage. Oxidants directly induce modification of genetic material, resulting in a number of different types of oxidized DNA bases.[48]

Phosphorylation of AhR and/or ARNT is essential for maximal AhR-mediated gene expression. The phosphorylation of tyrosine residues of AhR is important for DNA binding and treatment with tyrosine kinase inhibitors reduces AhR-mediated regulation of *CYP1A1*.[49] Moreover, phosphorylation of serine and threonine residues in the AhR transactivation domain is required for maximal transcriptional activity of AhR,[50] perhaps through promoting interactions with coregulator proteins.

More recently, AhR has been shown to have an important developmental role in vascular remodelling of the liver and eye, as well as an important role in neural development.[19,51] In the liver, AhR is essential for the proper postnatal closure of the ductus venosus, decreased hepatic perfusion, abnormal vasculature in the corneal limbus and cardiac hypertrophy.[52,53] Studies in recombinant animal models show that the adaptive, toxic and developmental pathways share similar upstream signalling events, including AHR activation, formation of the AhR/ARNT heterodimer, binding of the heterodimer complex to XREs and changes in gene expression.[20,54,55]

7.3 Dioxins, Furans and Coplanar PCBs

PCDDs and PCDFs are byproducts of combustion and of various industrial processes, and are widely present in the environment (Figure 7.3). They are released into the air during combustion processes such as industrial and municipal/household waste incineration, metal recycling and smelting operations and burning of fuels such as wood, coal, gasoline or oil.[9] They are also formed during forest fires and volcanic eruptions. PCDDs and PCDFs are two classes of chemicals that consist of 75 and 135 related compounds, respectively. Of the 210 PCDDs and PCDFs, only 17 are of toxicological concern. PCBs were manufactured in the past for a variety of industrial uses, notably as electrical insulators, dielectric fluids and specialized hydraulic fluids. There are 209 possible PCB congeners with 1 to 10 chlorine atoms on the biphenyl rings.

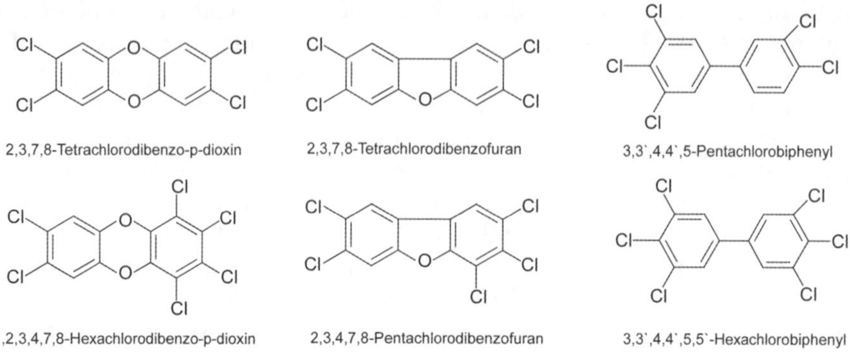

2,3,7,8-Tetrachlorodibenzo-p-dioxin   2,3,7,8-Tetrachlorodibenzofuran   3,3`,4,4`,5-Pentachlorobiphenyl

1,2,3,4,7,8-Hexachlorodibenzo-p-dioxin   2,3,4,7,8-Pentachlorodibenzofuran   3,3`,4,4`,5,5`-Hexachlorobiphenyl

Figure 7.3 Chemical structures of AhR-activating polychlorinated dibenzo-*p*-dioxins, polychlorinated dibenzofurans and polychlorinated biphenyls.

Congeners that have chlorine atoms only in *ortho* positions are non-coplanar due to steric hindrance, whereas those with chlorines only in *meta* and/or *para* positions are coplanar. Coplanar PCBs are ligands for AhR, but only 12 show toxicological properties similar to TCDD.[56] PCBs were marketed as mixtures of congeners known as Aroclors. Aroclor names reflect the number of carbons in the biphenyl rings and the percent chlorine (by weight) of the mixture (*e.g.* Aroclor 1242 is 42% chlorine by weight), with the more chlorinated mixtures generally being the most persistent and toxic. Most countries banned manufacture and use of PCBs in the 1970s; however, improper handling of past PCBs and disposal of contaminated equipment are still an environmental source of these compounds. Human populations show striking accumulations of PCBs, with reported levels of > 1700 ng g^{-1} lipid in human breast milk samples[57] and 3105 ng g^{-1} lipid in serum from residents residing near highly contaminated sites.[58] PCDDs, TCDFs and PCBs are lipophilic compounds that persist in the environment, accumulate in food chains and consequently end up in fatty tissues of animals and humans.

Although dioxin-like compounds are notoriously stable in the environment, the expression pattern of the CYP1A family has been reported to protect against dioxin toxicity in rodents. This enzyme metabolizes estrogen, uroporphyrinogen and melatonin.[59] In human populations there is > 60-fold variation in hepatic CYP1A2 levels.[60] Interestingly, the CYP1A2 protein sequesters planar AhR ligands[61] and the offspring of *Cyp1a2(-/-)* mice are ~ 6-fold more sensitive to dioxin-induced cleft palate, hydronephrosis and embryolethality, compared with *Cyp1a2(+/+)* mice. This effect is due to greater amounts of TCDD reaching the *Cyp1a2(-/-)* fetuses.[62]

7.3.1 Adverse Health Human Effects of Dioxin-like Compounds

Studies show that dioxin-like compounds have the potential to produce a wide range of effects on animals and humans. Some of these toxic responses include disruption of multiple endocrine pathways, teratogenesis, tumorigenesis, immunological dysfunction, neurobehavioral alterations, a general wasting syndrome and impaired reproduction.[11,63–72] The most commonly observed toxicity in highly exposed humans is chloracne, a pathological response of the skin. This condition is characterized by metaplasia and hyperkeratinization of the ducts of the sebaceous gland, resulting in comedone formation, as well as hyperproliferation and differentiation of the squamous epithelium.[73] Other adverse health effects of dioxin exposure in humans include cardiovascular disease, diabetes, cancer, endometriosis, early menopause, altered hormone levels, immunosuppression, altered growth factor signalling and altered metabolism.[9] Meta-analysis of data from three cohorts occupationally exposed to TCDD and related compounds found a statistically significant ($p = 0.02$) trend in total cancer mortality with increasing dioxin exposure,[74] including soft-tissue sarcomas, lung and liver cancers and breast cancer.[8,75,76]

In utero exposure of laboratory animals to dioxin-like compounds, including coplanar PCBs, can result in birth defects, such as cleft palate, hydronephrosis and immunosuppression.[9] Exposure during critical stages of development is also suspected to be associated with altered neurobehaviour, psychomotor function, cognition and development of reproductive organs. These findings are consistent with the ability of dioxin to alter behaviour in laboratory animals[77,78] and decreased the ability to learn discrimination tasks in non-human primates.[79] TCDD is known to directly or indirectly alter the levels and/or activity of other growth factors and hormones, such as estrogen, testosterone, thyroid hormone and gonadotropin-releasing hormone,[13] and may affect growth patterns in cells/tissues, leading to adverse outcomes. Developmental exposures to dioxin pose the greatest concern. High paternal exposure to dioxin has been reported to alter the sex ratio of offspring, with significantly more girls than boys being born.[80] In another study, men exposed to dioxin in their youth prior to puberty during the environmental accident in Seveso, Italy (see Section 9.5), exhibited reduced sperm count and motility as adults.[81] Higher exposures are found in nursing infants during periods of breastfeeding, causing worldwide concern and the re-evaluation of the presence of dioxin-related compounds in human milk.

7.4 Toxic Equivalent Factors

Exposure to PCDDs, PCDFs and coplanar PCBs occurs as mixture of congeners, with toxic effects similar to those of TCDD. This finding led to the development of the concept of toxic equivalent factors (TEFs). TEFs are based on the assumption that TCDD and other HAHs and PCBs have a common mechanism of action, *e.g.* activation of the AhR signalling pathway. The toxic potency of a particular congener, as described in studies *in vivo* and *in vitro*, is expressed relative to TCDD as a reference compound (the TEF for TCDD is arbitrarily set at 1; see Table 7.1). This assumption allows consideration of a broader range of data on toxicity, particularly in the case of human poisoning incidents. The benefit of this

Table 7.1 Compounds considered and the toxic equivalency factor assigned by WHO.[58]

Compound	Abbreviation	Toxic equivalency factor
Polychlorinated dibenzodioxins		
2,3,7,8-Tetrachlorodibenzo-*p*-dioxin	TCDD	1
1,2,3,7,8-Pentachlorodibenzodioxin	1,2,3,7,8-PeCDD	1
1,2,3,4,7,8-Hexachlorodibenzodioxin	1,2,3,4,7,8-HxCDD	0.1
Polychlorinated dibenzofurans		
2,3,7,8-Tetrachlorodibenzofuran	2,3,7,8-TCDF	0.1
1,2,3,7,8-Pentachlorodibenzofuran	1,2,3,7,8-PeCDF	0.05
2,3,4,7,8-Pentachlorodibenzofuran	2,3,4,7,8-PeCDF	0.3
Coplanar polychlorinated biphenyls		
3,3′,4,4′-Tetrachlorobiphenyl (polychlorinated biphenyl #77)	3,3′,4,4′-TCB	0.0001
3,3′,4,4′,5-Pentachlorobiphenyl (polychlorinated biphenyl #126)	3,3′,4,4′,5-PeCB	0.1
3,3′,4,4′,5,5′-Hexachlorobiphenyl (polychlorinated biphenyl #169)	3,3′,4,4′,5,5′-HxCB	0.01

assumption is that it allows data on the HAHs and PCBs to be summarized in a single description. The TEF approach relates the toxicity of all chemicals in the series to that of TCDD, for which most toxicological and epidemiological information is available. Known TEFs for the individual dioxins or dioxin-like components and their concentrations in the mixture can be utilized to determine a total toxic equivalency (TEQ) for the mixture. The TEQ is calculated as the sum of the individual products of the TEF and the concentration of each compound.[82] Many uncertainties exist in using the TEF approach for human risk assessment, including the fact that other compounds present in the environment may affect the biological response through AhR activation masking the "true" toxic equivalents of food. Toxic equivalence also is assumed to be additive in all cases, despite evidence that the effects of some compounds in environmental mixtures are synergistic or antagonistic.[57] The TEF concept does not apply to halogenated compounds other than PCDDs, PCDFs and PCBs which show AhR-dependent toxicity, including brominated analogues of PCDDs and PCDFs.[57] However, pragmatically it is the most feasible approach available. The WHO used the TEF concept in 1998[83] to establish a tolerable daily intake (TDI) of 1–4 pg kg^{-1} bodyweight (bw), which was applied to the toxic equivalents of PCDDs, PCDFs and coplanar PCBs. The mean daily intake of dioxin-like compounds varies among countries. Studies show that the average daily intake of dioxin-like compounds is 0.5–2 pg kg^{-1} bw of "international" dioxin toxic equivalents.[83]

7.5 Sources of Human Exposure

Human exposure to dioxin-like compounds occurs through environmental, accidental and occupational contamination and dietary sources. The major pathway of human exposure is through contaminated food and predominantly from animal fat.[9,83,84] Contamination of food is caused by deposition of emissions of various sources (*i.e.* waste incineration, production of chemicals, metal industry) on farmland and a subsequent accumulation in the food chain, in which they accumulate within fat. Animal fat contamination is derived largely from animal feed, making it a potential control point for reduction of these compounds from entering the food chain. Other sources may include improper application of sewage sludge, flooding of pastures and waste effluents and certain types of food processing.[85]

7.5.1 PCDDs, PCDFs and Coplanar PCBs that Contaminate Food

The major pathway of human exposure is through contaminated food and predominantly from animal fat.[9,83,84] The sources of PCBs are very different from those of PCDDs and PCDFs, since there was a substantial commercial production of PCBs. Since the commercial production of PCBs began almost 70 years ago, it has been estimated that 1.5 million tons were produced.[86] Over this time, PCB-containing industrial discharges and the improper use and disposal of equipment containing PCBs have contributed to their widespread contamination. Because their manufacture and use was banned in most countries in the 1970s, the predominant source of PCBs is through past releases. In the past, the chemical industry was the main source of releases of PCDDs and PCDFs into the environment.[86] Today, the most significant releases of these compounds are from combustion processes.

Environmental contamination in animal feed is the most common mode by which PCDDs, PCDFs and coplanar PCBs enter into the food chain. Feed, farmed animals and food products may become contaminated in various ways, including deposition of emissions onto farmland, blending of feedstuffs with contaminated products, application of contaminated pesticides or disinfectants, contact with materials treated with wood preservatives, application of sewage sludge to fields, flooding of pastures, and contamination of water with wastewater and/or effluents. Widespread environmental contamination with PCDDs, PCDFs and coplanar PCBs remains after past releases. As the half-lives of some of these compounds in the environment are decades or longer, the environmental contamination is likely to persist for some time.

7.5.2 Accidental Human Exposure and Entry into the Food Chain

During the past few decades, heavy exposure to dioxin-like compounds has occurred in a number of isolated incidents. One of the well-studied examples of environmental releases is the exposure of the local population at Seveso, Italy.[87] In 1976, an explosion at a chemical plant in Seveso producing 2,4,5-trichlorophenol, an intermediate in the synthesis of the herbicide

2,4,5-trichlorophenoxyacetic acid (2,4,5-T). TCDD, a contaminate of 2,4,5-T, was released into the environment. Within weeks of the accident, community members experienced chloracne. Continued studies on the exposed population supports the potential of TCDD as a human carcinogen and as a risk factor in a number of human diseases.[81,88]

Heavy human exposure to dioxin-like compounds has also been caused by contaminated foods. In 1968 in Kyushu, Japan, a company's rice oil supplies were contaminated with PCBs and PCDFs. The contaminated oil was later sold for human consumption and as animal feed to livestock. Exposed individuals experienced chloracne, fatigue and altered reproductive and immune function, as well as developmental problems in children in what was referred to as Yusho disease.[89] A similar incident occurred in Taiwan in 1979, in which food-grade rice cooking oil was contaminated by dioxin-like compounds. This occurrence is referred to as Yucheng disease, with the exposures being remarkably similar in terms of the amount of total exposure to the contaminants and the number of people exposed as those in the Yusho incident. Both of these accidents involved exposure to concentrations of PCDDs, PCDFs or PCBs at least three to four orders of magnitude higher than those normally found in foods.

Other incidents of human exposure through contaminated foods have been reported. The use of a naturally contaminated feed additive (ball clay) led to elevated concentrations of dioxin in farmed catfish and poultry and the subsequent banning of ball clay in feed.[85] Moreover, a large international analysis of farmed *versus* wild salmon revealed that farmed fish had elevated levels of dioxin-like compounds compared to their wild counterparts.[90,91]

In Belgium, a mixture of PCBs contaminated with dioxins was accidentally added to a stock of recycled fat used in the production of animal feeds, resulting in a massive international recall of contaminated poultry, eggs, milk, meat and chocolate.[92] More recently, in 2008, all Irish pork products were removed from store shelves owing to known contamination of these animals with dioxin-like chemical contaminated feed. The successful reduction of emissions of PCDDs, PCDFs and coplanar PCBs into the environment in the 1970s, 1980s and 1990s has reduced our overall exposure levels to these compounds. Attention must now be placed on animal feed and pathways into feed that remain a vulnerable point by which these compounds enter the food supply.

7.6 Natural Dietary Compounds that Modulate AhR Action

In the previous sections we discussed the food contaminants that mediate their toxic effects through AhR. There are, however, a number of natural chemicals found in plants that modulate AhR function and many of them have been described as health promoting, disease preventing and cancer preventative agents. These compounds include indoles, flavonoids, polyphenolic compounds, carotinoids and retinoids.[16] The structures of some of these

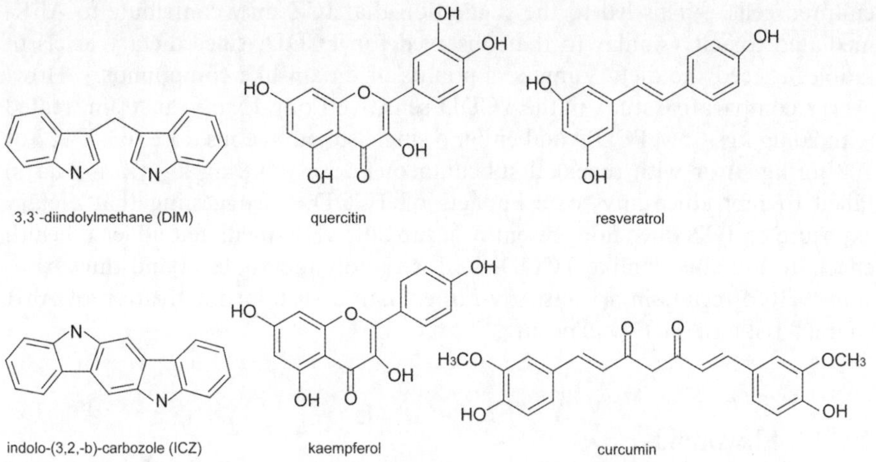

3,3`-diindolylmethane (DIM)   quercitin   resveratrol

indolo-(3,2,-b)-carbozole (ICZ)   kaempferol   curcumin

Figure 7.4 Chemical structures of typical dietary AhR modulating compound.

compounds are shown in Figure 7.4. These compounds bind to AhR and act as agonists or antagonists.[2] They also inhibit many of the drug metabolizing enzymes induced by AhR and act as anti-oxidants.[93] A discussion of the properties of selected phytochemicals and how they modulate AhR signalling is presented in the following sections.

7.6.1 Indoles

Several studies have described natural compounds that bind to AhR and either activate or inhibit its activity. The abilities of natural indole compounds to modulate AhR function were the earliest indications of the existence of natural AhR inducers.[94,95] Indole-3-carbinol (IC3) is a weak AhR activator found in cruciferous vegetables (*i.e.* broccoli, cauliflower and Brussels sprouts).[95] In the gastrointestinal tract, IC3 conjugates are hydrolyzed to many products, including the AhR agonists 3,3′-diindolylmethane (DIM) and indolo[3,2,-*b*]carbazole (ICZ). Dietary intake levels of DIM positively activate AhR; however, at these same doses, DIM has been reported to specifically activate estrogen receptor β.[96] Moreover, studies in estrogen receptor α (ERα)–positive MCF-7 human breast cancer cells show that DIM induces ERα-activated transcription of E2-responsive genes.[97] DIM is also a potent ligand-independent activator of ERα through protein kinase A and activates multiple signalling pathways, including those that involve nuclear factor-kappaB signalling, caspase activation, DNA repair and cytochrome P450 metabolism, as well as other cellular kinases.[98] It remains to be determined whether AhR mediates many of the above-mentioned properties of DIM.

ICZ has one of the highest affinities for AhR of any phytochemical identified to date (∼0.2–3.6 nM) and is a potent inducer of CYP1A1 expression in cultured cells.[95] This led to the suggestion that ICZ may contribute to AhR-mediated toxicity similar to that observed for TCDD, since dietary levels of indoles exceed the dietary intake amounts of dioxin-like compounds.[99] However, a comparative study in the TCDD-sensitive Long-Evans rat strain treated with 20 µg kg^{-1} bw TCDD and either a single subcutaneous dose (63.5, 127 or 508 µg kg^{-1}) or with repeated subcutaneous doses (508 µg kg^{-1} for 5 days) failed to reproduce any toxic impacts of TCDD,[100] suggesting that dietary exposure to ICZ does not present a hazard of AHR-mediated adverse health effects to humans. Unlike TCDD, ICZ is rapidly metabolized and thus exposure to ICZ results in a transitory rather than a sustained activation of AhR characteristic of TCDD exposure.[45]

7.6.2 Flavonoids

Flavonoids are a large class of naturally occurring compounds found in vegetables, fruits, nuts, tea and red wine, as well as in many dietary supplements and herbal remedies.[101] The average intake of flavonoids in the human diet has been estimated to be ~ 1 g d^{-1}.[102] However, specific flavonoids are marketed as nutritional supplements and as a result the intake of these compounds is likely to be higher. More than 8000 compounds with flavonoid structures that differ in the location of hydroxy and methoxy group substituents on the flavonoid ring structure have been identified.[103] Flavonoids are grouped into different structural classes, including flavones, flavonols, flavanones, isoflavones, anthocyanins and catechins. These classes of compounds exert a wide range of pharmacological properties, with their ability to act as cancer preventative agents being most extensively studied.[103] These cancer protective properties are attributed to a wide variety of mechanisms, including their ability to act as radical scavengers, alter the functions of modifying enzymes that activate or prevent the generation of carcinogens and cause changes in gene expression through the modulation of transcription factor activity.[93,103] Some flavones also exert antibacterial and anti-allergic activities, inhibit lipid peroxidation and act as metal chelators.[104–106]

Numerous flavonoids directly bind and activate or inhibit the AhR signalling pathway and represent the largest group of naturally occurring dietary AhR ligands.[16] Many of the natural flavones bind to AhR in the low µM range;[2] however, studies of human volunteers reveal that many flavones, including quercitin, are found at physiological concentrations that activate AhR.[107] Two synthetic flavonoids, β-naphthoflavone and α-naphthoflavone, are commonly used as complete agonists and partial antagonists of AhR, respectively.[108] Although the vast majority of flavonoids are AhR antagonists, there are examples of AhR agonists among the different classes of flavonoids.[2] The most abundant dietary flavonoids, quercitin and kaempferol, are both AhR ligands but they exert very different effects on AhR-dependent regulation of CYP1A1.[15] Quercetin induces time- and dose-dependent increases in CYP1A1 mRNA levels in different human cell line models.[15] However, antagonistic effects of quercetin have also been reported, suggesting that phytochemicals exhibit substantial cell context-dependent AhR agonist and antagonist activities.[109] Similar to quercetin, galangin, a flavonol found in honey, exerts agonist and antagonistic effects on AhR.[110] Kaempferol alone has no effect on CYP1A1 mRNA levels, but antagonizes TCDD-induced CYP1A1 mRNA levels.[15] Recent studies have shown that kaempferol prevents TCDD-dependent recruitment of AhR to its target genes, blocking transcriptional activation by AhR.[111] Teas, such as green tea, are rich sources of catechins, and some of these compounds have been shown to suppress AhR activity.[103] The most effective catechin is (–)-epigallocatechin-3-gallate (EGCG).[112] EGCG suppress AhR action through its ability to bind to Hsp90. Tea flavins in black tea have also been reported to antagonize AhR. Many flavonoids also inhibit the enzymatic activities of some phase I and II drug metabolizing enzymes, including CYP1A1, CYP1B1, CYP1A2, CYP3A4 and sulfotransferase 1E1 (SULT1E1) and SULT1A2.[103,113] Their ability to specifically modulate these enzymes is thought to significantly contribute to their protective effects by preventing the metabolic activation of pro-carcinogens.[103]

7.6.3 Resveratrol

Resveratrol (*trans*-3,4′,5-trihydroxystilbene) is a polyphenol present in grape skin, wine, peanuts and berries.[114] Resveratrol has gained intense interest due to its anticarcinogenic, cardioprotective and anti-aging properties.[114,115] These properties have been attributed to its ability to act as an anti-oxidant, anti-inflammatory agent and pro-apoptotic agent.[116] The term the "French paradox", coined in 1992, refers to the relatively low incidence of cardiovascular disease in the French population, despite a relatively high consumption of saturated fats.[117] Some researchers suggest that the low incidences in cardiovascular disease are due to the high relative consumption of red wine.[117] Resveratrol is one of many possible polyphenolic compounds in red wine that contribute to preventing the development and progression of cardiovascular disease.[118]

Resveratrol suppresses the AhR-dependent transcription through direct binding to AhR and prevents activated AhR from being recruited to and interacting with XREs.[111,119,120] Similar to other phytochemicals, resveratrol not only interacts with AhR but modulates other cellular signalling pathways, including acting as an agonist for the estrogen receptor (ER).[121] Resveratrol inhibits the enzymatic activities of CYP1A1 and CYP1A2 through suppressing their induction by AhR.[122] Although resveratrol shows many beneficial effects, its bioavailability is very low and it is known to be rapidly metabolized into sulfate and glucuronide conjugates, with resveratrol glucuronide predominantly circulating in human plasma.[116,123] The phase II conjugation of resveratrol significantly reduces its bioavailability *in vivo*, which might weaken the *in vitro* anticancer, anti-inflammatory and anti-atherogenic activities attributed to it. Therefore, metabolism and bioavailability should be carefully considered in future investigations. However, resveratrol administered by subcutaneous injection to female Sprague-Dawley rats completely prevented benzo[*a*]pyrene-dependent induction of CYP1A1 protein expression in several tissues.[124] These findings demonstrate the effectiveness of resveratrol at inhibiting AhR-mediated activity *in vivo* and show that phase II metabolic products of certain phytochemicals retain biological activity.

7.6.4   Curcumins

Curcumin is a polyphenol agent present in the plant *Curcuma longa*. Curcumin, in the form of the herbal powder turmeric, has been used for centuries as an anti-inflammatory remedy in Asian medicine. Turmeric has been widely consumed traditionally for gastrointestinal and respiratory problems such as chronic diarrhea, flatulence, cough, cold, asthma and throat irritations.[125] Curcumin has demonstrated a variety of *in vitro* cancer-inhibitory effects, including inhibition of mitogenic signalling, suppression of cell growth and induction of apoptosis.[126,127] Curcumin is an AhR antagonist and prevents TCDD-mediated induction of CYP1A1 and CYP1B1 mRNA and protein levels, as well as preventing the anchorage-independent growth induced by TCDD of normal human embryonic kidney cells and normal prostate cells.[128,129] It also inhibits CYP1A1-mediated bioactivation of the tobacco-associated carcinogen benzo[*a*]pyrene in human oral keratinocytes and human oral mucosal tissues.[130] Moreover, curcumin inhibits benzo[*a*]pyrene-induced stomach cancer in rodents, and competitively inhibits bioactivation of dimethylbenzanthracene, suggesting that this phytochemical not only inhibits AhR induction of CYP1A1 but that it also functions as a CYP1A1 substrate.[129,131] Overall, these data suggest that the chemopreventive efficacy of curcumin is in part due to its modulation of cellular carcinogen metabolism and metabolizing enzymes, some of which are regulated by AhR.

7.6.5   Carotenoids and Retinoids

Numerous carotenoids and retinoids have been reported to competitively bind to AhR and modulate AhR-dependent gene expression.[14,132] The carotenoids canthaxanthin, astaxanthin and β-apo-8′-carotenal induced CYP1A1 and CYP1A2 activity in rodents, whereas β-carotene and lutein had no effect.[133–135] Subsequent studies showed that lutein suppresses TCDD-dependent transformation of AhR in rat hepatic cytosol and in human immortalized cell lines.[112]

The synthetic retinoids AGN193109, AGN190730 and AGN192837, which activate the retinoic acid receptor/retinoid X receptor pathway, also bind to AhR and activate the AhR signalling pathway.[136] However, many natural retinoids suppress AhR action *in vitro* and *in vivo*. Treatment with vitamin A reduced the severity of TCDD-induced liver damage and reduced CYP1A1 expression compared to TCDD-treated mice, indicating that supplementation of vitamin A might attenuate the liver damage caused by TCDD.[137] Retinoic acid suppresses 3-methylcholanthrene-induced CYP1A1 expression in Caco-2 human epithelial colorectal adenocarcinoma cells. Vitamin A, all *trans*-retinol, all *trans*-retinal and all *trans*-retinoic acid inhibit CYP1A1 enzymatic activity, leading the authors to suggest that the ingestion of enough amounts of vitamins A from foods might lead to the inhibition of the activity of P4501A1, which is known to be induced by smoking, pharmaceutical drugs and environmental pollutants like dioxins.[138] Collectively, the current data suggest that synthetic retinoids act as agonists, while natural retinoids act as antagonists for the AhR.

7.6.6   Protective Effects of Dietary AhR Modulating Compounds

In the previous sections we discussed a few of the numerous phytochemicals that are reported to modulate AhR signalling. Owing to space constraint, only a few classes of AhR modulating compounds were discussed and the reader is referred to recent reviews for further information.[2,16,103] Based on the wide distribution of AhR modulating compounds in dietary foods, it is not surprising the many crude food extracts possess AhR modulating activity.[139] Many of the plant-derived materials contain AhR ligands or contain compounds that are readily converted into AhR ligands, and as such represent the largest natural source of AhR-modulating compounds. The majority of these ligands suppress AhR activity or inhibit many of the drug metabolizing enzymes regulated by AhR; thus it remains possible that dietary intake of AhR-modulating compounds might prevent the toxicities associated with exposure to AhR activating dioxin-like compounds. Phytochemicals suppress AhR-dependent activation, but also inhibit many of the drug metabolizing enzymes regulated by AhR, preventing the generation of toxic and potentially carcinogenic metabolites. Current research is largely based on studies in immortalized animal and to some extent human cell lines, with little evidence that phytochemicals prevent dioxin- or PAH-induced toxicity in rodent models. There is increasing evidence that rodent and human AhRs exhibit significant functional differences and distinct ligand preferences,[140,141] suggesting that data generated in cell and animal models might not be readily applicable to humans.

7.7   Conclusions

The AhR is a highly conserved ligand-activated transcription factor that plays a key role in growth and development and in the adaptive response to xenobiotics *via* the regulation of drug metabolizing enzymes. AhR also mediates the toxic response of TCDD and other dioxin-like environmental contaminants. Dioxin-like compounds known to activate AhR have been associated with a number of adverse health effects, including cancer. It is estimated that 90% of human exposure to dioxin-like compounds occurs through the intake of contaminated food. AhR activity is modulated by numerous phytochemicals that act as agonists or antagonists, with the majority of them acting as AhR antagonists at dietary levels. Many phytochemicals target numerous cellular and other nuclear receptor pathways, as well as inhibiting the catalytic activity of a wide variety of drug metabolizing enzymes. Dietary intake of AhR-modulating phytochemicals might suppress the toxicity elicited by dioxin-like compounds. However, the bioavailability of many phytochemicals is low and these compounds are susceptible to rapid conjugation and metabolism. This is in contrast to dioxin-like compounds that have relatively long half-lives (7.5 years) in humans and accumulate in adipose tissue. Therefore differences in the metabolism of phytochemicals need to be considered in determining the impact that these compounds have on AhR-mediated signalling in humans.

References

1. B. E. McIntosh, J. B. Hogenesch and C. A. Bradfield, *Annu. Rev. Physiol.*, 2010, **72**, 625.
2. M. S. Denison and S. R. Nagy, *Annu. Rev. Pharmacol. Toxicol.*, 2003, **43**, 309.
3. J. Song, M. Clagett-Dame, R. E. Peterson, M. E. Hahn, W. M. Westler, R. R. Sicinski and H. F. DeLuca, *Proc. Natl. Acad. Sci. USA*, 2002, **99**, 14694.
4. J. F. Savouret, M. Antenos, M. Quesne, J. Xu, E. Milgrom and R. F. Casper, *J. Biol. Chem.*, 2001, **276**, 3054.
5. A. Schecter and J. R. Olson, *Chemosphere*, 1997, **34**, 1569.
6. IARC working group, *Evaluation of Carcinogenic Risks to Humans*, IARC Monographs, IARC Press, Lyon, 1997, vol. 69, p. 33.
7. S. Safe, *Toxicol. Lett.*, 2001, **120**, 1.
8. K. Steenland, P. Bertazzi, A. Baccarelli and M. Kogevinas, *Environ. Health Perspect.*, 2004, **112**, 1265.
9. U.S. EPA, 2004, National Academy of Sciences (NAS) Review Draft 200. EPA/600/P-00/001Cb.
10. L. S. Birnbaum, *Environ. Health Perspect.*, 1994, **102**(suppl. 9), 157.
11. L. S. Birnbaum, *Toxicol. Lett.*, 1995, **82/83**, 743.
12. A. K. Liem, P. Furst and C. Rappe, *Food Addit. Contam.*, 2000, **17**, 241.
13. S. H. Safe, *J. Anim. Sci.*, 1998, **76**, 134.
14. S. Gradelet, P. Astorg, J. Leclerc, J. Chevalier, M. F. Vernevaut and M. H. Siess, *Xenobiotica*, 1996, **26**, 49.
15. H. P. Ciolino, P. J. Daschner and G. C. Yeh, *Biochem. J.*, 1999, **340** (part 3), 715.
16. H. Ashida, S. Nishiumi and I. Fukuda, *Expert Opin. Drug Metab. Toxicol.*, 2008, **4**, 1429.
17. Y. Z. Gu, J. B. Hogenesch and C. A. Bradfield, *Annu. Rev. Pharmacol. Toxicol.*, 2000, **40**, 519.
18. M. E. Hahn, *Chem. Biol. Interact.*, 2002, **141**, 131.
19. J. A. Walisser, M. K. Bunger, E. Glover, E. B. Harstad and C. A. Bradfield, *J. Biol. Chem.*, 2004, **279**, 16326.

20. J. V. Schmidt, G. H. Su, J. K. Reddy, M. C. Simon and C. A. Bradfield, *Proc. Natl. Acad. Sci. USA*, 1996, **93**, 6731.
21. J. Mimura, K. Yamashita, K. Nakamura, M. Morita, T. N. Takagi, K. Nakao, M. Ema, K. Sogawa, M. Yasuda, M. Katsuki and Y. Fujii-Kuriyama, *Genes Cells*, 1997, **2**, 645.
22. T. Baba, J. Mimura, N. Nakamura, N. Harada, M. Yamamoto, K. Morohashi and Y. Fujii-Kuriyama, *Mol. Cell. Biol.*, 2005, **25**, 10040.
23. P. M. Fernandez-Salguero, D. M. Hilbert, S. Rudikoff, J. M. Ward and F. J. Gonzalez, *Toxicol. Appl. Pharmacol.*, 1996, **140**, 173.
24. J. Micka, A. Milatovich, A. Menon, G. A. Grabowski, A. Puga and D. W. Nebert, *Pharmacogenetics*, 1997, **7**, 95.
25. M. E. Hahn, *Comp. Biochem. Physiol. C: Pharmacol. Toxicol. Endocrinol.*, 1998, **121**, 23.
26. L. S. Birnbaum, M. M. McDonald, P. C. Blair, A. M. Clark and M. W. Harris, *Fundam. Appl. Toxicol.*, 1990, **15**, 186.
27. A. Poland, D. Palen and E. Glover, *Mol. Pharmacol.*, 1994, **46**, 915.
28. R. Pohjanvirta, J. M. Wong, W. Li, P. A. Harper, J. Tuomisto and A. B. Okey, *Mol. Pharmacol.*, 1998, **54**, 86.
29. I. D. Moffat, P. C. Boutros, H. Chen, A. B. Okey and R. Pohjanvirta, *BMC Genomics*, 2010, **11**, 263.
30. J. P. Whitlock, Jr., *Annu. Rev. Pharmacol. Toxicol.*, 1999, **39**, 103.
31. M. Denis, S. Cuthill, A. C. Wikstrom, L. Poellinger and J. A. Gustafsson, *Biochem. Biophys. Res. Commun.*, 1988, **155**, 801.
32. G. H. Perdew, *J. Biol. Chem.*, 1988, **263**, 13802.
33. L. A. Carver and C. A. Bradfield, *J. Biol. Chem.*, 1997, **272**, 11452.
34. R. S. Pollenz, *Chem. Biol. Interact.*, 2002, **141**, 41.
35. O. Hankinson, *Arch. Biochem. Biophys.*, 2005, **433**, 379.
36. O. Hankinson, *Annu. Rev. Pharmacol. Toxicol.*, 1995, **35**, 307.
37. J. L. Marlowe, E. S. Knudsen, S. Schwemberger and A. Puga, *J. Biol. Chem.*, 2004, **279**, 29013.
38. G. Huang and C. J. Elferink, *Mol. Pharmacol.*, 2005, **67**, 88.
39. S. Ahmed, E. Valen, A. Sandelin and J. Matthews, *Toxicol. Sci.*, 2009, **111**, 254.
40. M. Abdelrahim, R. Smith, 3rd and S. Safe, *Mol. Pharmacol.*, 2003, **63**, 1373.
41. J. Mimura, M. Ema, K. Sogawa and Y. Fujii-Kuriyama, *Genes Dev.*, 1999, **13**, 20.
42. J. Mimura and Y. Fujii-Kuriyama, *Biochim. Biophys. Acta*, 2003, **1619**, 263.
43. Y. Kanno, Y. Takane, Y. Takizawa and Y. Inouye, *Mol. Cell. Endocrinol.*, 2008, **291**, 87.
44. J. M. Pascussi, S. Gerbal-Chaloin, C. Duret, M. Daujat-Chavanieu, M. J. Vilarem and P. Maurel, *Annu. Rev. Pharmacol. Toxicol.*, 2008, **48**, 1.
45. K. W. Bock and C. Kohle, *Biochem. Pharmacol.*, 2006, **72**, 393.
46. A. H. Conney, *Cancer Res.*, 2003, **63**, 7005.
47. J. H. Park, D. Mangal, A. J. Frey, R. G. Harvey, I. A. Blair and T. M. Penning, *J. Biol. Chem.*, 2009, **284**, 29725.

48. R. De Bont and N. van Larebeke, *Mutagenesis*, 2004, **19**, 169.
49. I. Pongratz, P. E. Stromstedt, G. G. Mason and L. Poellinger, *J. Biol. Chem.*, 1991, **266**, 16813.
50. Y. H. Chen and R. H. Tukey, *J. Biol. Chem.*, 1996, **271**, 26261.
51. G. P. Lahvis and C. A. Bradfield, *Biochem. Pharmacol.*, 1998, **56**, 781.
52. G. P. Lahvis, R. W. Pyzalski, E. Glover, H. C. Pitot, M. K. McElwee and C. A. Bradfield, *Mol. Pharmacol.*, 2005, **67**, 714.
53. E. B. Harstad, C. A. Guite, T. L. Thomae and C. A. Bradfield, *Mol. Pharmacol.*, 2006, **69**, 1534.
54. M. K. Bunger, E. Glover, S. M. Moran, J. A. Walisser, G. P. Lahvis, E. L. Hsu and C. A. Bradfield, *Toxicol. Sci.*, 2008, **106**, 83.
55. M. K. Bunger, S. M. Moran, E. Glover, T. L. Thomae, G. P. Lahvis, B. C. Lin and C. A. Bradfield, *J. Biol. Chem.*, 2003, **278**, 17767.
56. J. C. Larsen, *Mol. Nutr. Food Res.*, 2006, **50**, 885.
57. M. Van den Berg, L. S. Birnbaum, M. Denison, M. De Vito, W. Farland, M. Feeley, H. Fiedler, H. Hakansson, A. Hanberg, L. Haws, M. Rose, S. Safe, D. Schrenk, C. Tohyama, A. Tritscher, J. Tuomisto, M. Tysklind, N. Walker and R. E. Peterson, *Toxicol. Sci.*, 2006, **93**, 223.
58. J. Petrik, B. Drobna, M. Pavuk, S. Jursa, S. Wimmerova and J. Chovancova, *Chemosphere*, 2006, **65**, 410.
59. B. Wang and S. F. Zhou, *Curr. Med. Chem.*, 2009, **16**, 4066.
60. D. W. Nebert, T. P. Dalton, A. B. Okey and F. J. Gonzalez, *J. Biol. Chem.*, 2004, **279**, 23847.
61. H. Hakk, J. J. Diliberto and L. S. Birnbaum, *Toxicol. Appl. Pharmacol.*, 2009, **241**, 119.
62. N. Dragin, T. P. Dalton, M. L. Miller, H. G. Shertzer and D. W. Nebert, *J. Biol. Chem.*, 2006, **281**, 18591.
63. R. J. Kociba, D. G. Keyes, J. E. Beyer, R. M. Carreon, C. E. Wade, D. A. Dittenber, R. P. Kalnins, L. E. Frauson, C. N. Park, S. D. Barnard, R. A. Hummel and C. G. Humiston, *Toxicol. Appl. Pharmacol.*, 1978, **46**, 279.
64. L. E. Gray, J. S. Ostby and W. R. Kelce, *Toxicol. Appl. Pharmacol.*, 1997, **146**, 11.
65. L. E. Gray, Jr., J. S. Ostby and W. R. Kelce, *Toxicol. Appl. Pharmacol.*, 1994, **129**, 46.
66. J. T. Tuomisto, R. Pohjanvirta, M. Unkila and J. Tuomisto, *Pharmacol. Biochem. Behav.*, 1999, **62**, 735.
67. J. E. Chastain, Jr. and T. L. Pazdernik, *Int. J. Immunopharmacol.*, 1985, **7**, 849.
68. M. G. Wade, W. G. Foster, E. V. Younglai, A. McMahon, K. Leingartner, A. Yagminas, D. Blakey, M. Fournier, D. Desaulniers and C. L. Hughes, *Toxicol. Sci.*, 2002, **67**, 131.
69. H. M. Miettinen, P. Pulkkinen, T. Jamsa, J. Koistinen, U. Simanainen, J. Tuomisto, J. Tuukkanen and M. Viluksela, *Toxicol. Sci.*, 2005, **85**, 1003.
70. A. Poland and J. C. Knutson, *Annu. Rev. Pharmacol. Toxicol.*, 1982, **22**, 517.
71. A. Schecter, L. Birnbaum, J. J. Ryan and J. D. Constable, *Environ. Res.*, 2006, **101**, 419.

72. L. S. Birnbaum, *Environ. Health Perspect.*, 1994, **102**, 676.

73. A. A. Panteleyev and D. R. Bickers, *Exp. Dermatol.*, 2006, **15**, 705.

74. K. S. Crump, R. Canady and M. Kogevinas, *Environ. Health Perspect.*, 2003, **111**, 681.

75. H. Kang, F. M. Enzinger, P. Breslin, M. Feil, Y. Lee and B. Shepard, *J. Natl. Cancer Inst.*, 1987, **79**, 693.

76. M. Warner, B. Eskenazi, P. Mocarelli, P. M. Gerthoux, S. Samuels, L. Needham, D. Patterson and P. Brambilla, *Environ. Health Perspect.*, 2002, **110**, 625.

77. R. Hojo, M. Kakeyama, Y. Kurokawa, Y. Aoki, J. Yonemoto and C. Tohyama, *Environ. Health Prev. Med.*, 2008, **13**, 169.

78. R. Hojo, S. Stern, G. Zareba, V. P. Markowski, C. Cox, J. T. Kost and B. Weiss, *Environ. Health Perspect.*, 2002, **110**, 247.

79. S. L. Schantz and R. E. Bowman, *Neurotoxicol. Teratol.*, 1989, **11**, 13.

80. P. Mocarelli, P. M. Gerthoux, E. Ferrari, D. G. Patterson, Jr., S. M. Kieszak, P. Brambilla, N. Vincoli, S. Signorini, P. Tramacere, V. Carreri, E. J. Sampson, W. E. Turner and L. L. Needham, *Lancet*, 2000, **355**, 1858.

81. P. Mocarelli, P. M. Gerthoux, D. G. Patterson, Jr., S. Milani, G. Limonta, M. Bertona, S. Signorini, P. Tramacere, L. Colombo, C. Crespi, P. Brambilla, C. Sarto, V. Carreri, E. J. Sampson, W. E. Turner and L. L. Needham, *Environ. Health Perspect.*, 2008, **116**, 70.

82. M. Van den Berg, L. Birnbaum, A. T. Bosveld, B. Brunstrom, P. Cook, M. Feeley, J. P. Giesy, A. Hanberg, R. Hasegawa, S. W. Kennedy, T. Kubiak, J. C. Larsen, F. X. van Leeuwen, A. K. Liem, C. Nolt, R. E. Peterson, L. Poellinger, S. Safe, D. Schrenk, D. Tillitt, M. Tysklind, M. Younes, F. Waern and T. Zacharewski, *Environ. Health Perspect.*, 1998, **106**, 775.

83. F. X. van Leeuwen, M. Feeley, D. Schrenk, J. C. Larsen, W. Farland and M. Younes, *Chemosphere*, 2000, **40**, 1095.

84. J. Huwe, D. Pagan-Rodriguez, N. Abdelmajid, N. Clinch, D. Gordon, J. Holterman, E. Zaki, M. Lorentzsen and K. Dearfield, *J. Agric. Food Chem.*, 2009, **57**, 11194.

85. C. Rappe, S. Bergek, H. Fiedler and K. R. Cooper, *Chemosphere*, 1998, **36**, 2705.

86. S. S. White and L. S. Birnbaum, *J. Environ. Sci. Health C*, 2009, **27**, 197.

87. F. Pocchiari, V. Silano and A. Zampieri, *Ann. N. Y. Acad. Sci.*, 1979, **320**, 311.

88. P. A. Bertazzi, D. Consonni, S. Bachetti, M. Rubagotti, A. Baccarelli, C. Zocchetti and A. C. Pesatori, *Am. J. Epidemiol.*, 2001, **153**, 1031.

89. Y. Aoki, *Environ. Res.*, 2001, **86**, 2.

90. J. A. Foran, D. O. Carpenter, M. C. Hamilton, B. A. Knuth and S. J. Schwager, *Environ. Health Perspect.*, 2005, **113**, 552.

91. R. A. Hites, J. A. Foran, D. O. Carpenter, M. C. Hamilton, B. A. Knuth and S. J. Schwager, *Science*, 2004, **303**, 226.

92. A. Bernard, F. Broeckaert, G. De Poorter, A. De Cock, C. Hermans, C. Saegerman and G. Houins, *Environ. Res.*, 2002, **88**, 1.

93. P. Hodek, P. Trefil and M. Stiborova, *Chem. Biol. Interact.*, 2002, **139**, 1.
94. M. Gillner, J. Bergman, C. Cambillau, M. Alexandersson, B. Fernstrom and J. A. Gustafsson, *Mol. Pharmacol.*, 1993, **44**, 336.
95. L. F. Bjeldanes, J. Y. Kim, K. R. Grose, J. C. Bartholomew and C. A. Bradfield, *Proc. Natl. Acad. Sci. USA*, 1991, **88**, 9543.
96. O. I. Vivar, E. F. Saunier, D. C. Leitman, G. L. Firestone and L. F. Bjeldanes, *Endocrinology*, 2010, **151**, 1662.
97. J. E. Riby, G. H. Chang, G. L. Firestone and L. F. Bjeldanes, *Biochem. Pharmacol.*, 2000, **60**, 167.
98. J. R. Weng, C. H. Tsai, S. K. Kulp and C. S. Chen, *Cancer Lett.*, 2008, **262**, 153.
99. S. Safe, F. Wang, W. Porter, R. Duan and A. McDougal, *Toxicol. Lett.*, 1998, **102/103**, 343.
100. R. Pohjanvirta, M. Korkalainen, J. McGuire, U. Simanainen, R. Juvonen, J. T. Tuomisto, M. Unkila, M. Viluksela, J. Bergman, L. Poellinger and J. Tuomisto, *Food Chem. Toxicol.*, 2002, **40**, 1023.
101. R. J. Nijveldt, E. van Nood, D. E. van Hoorn, P. G. Boelens, K. van Norren and P. A. van Leeuwen, *Am. J. Clin. Nutr.*, 2001, **74**, 418.
102. A. Scalbert and G. Williamson, *J. Nutr.*, 2000, **130**, 2073S.
103. Y. J. Moon, X. Wang and M. E. Morris, *Toxicol. In vitro*, 2006, **20**, 187.
104. E. Middleton, Jr., *Adv. Exp. Med. Biol.*, 1998, **439**, 175.
105. J. D. Lambert and R. J. Elias, *Arch. Biochem. Biophys.*, 2009, **501**, 65.
106. H. Yoshida, T. Ishikawa, H. Hosoai, M. Suzukawa, M. Ayaori, T. Hisada, S. Sawada, A. Yonemura, K. Higashi, T. Ito, K. Nakajima, T. Yamashita, K. Tomiyasu, M. Nishiwaki, F. Ohsuzu and H. Nakamura, *Biochem. Pharmacol.*, 1999, **58**, 1695.
107. G. Paganga and C. A. Rice-Evans, *FEBS Lett.*, 1997, **401**, 78.
108. B. Wihlen, S. Ahmed, J. Inzunza and J. Matthews, *Mol. Cancer Res.*, 2009, **7**, 977.
109. S. Zhang, C. Qin and S. H. Safe, *Environ. Health Perspect.*, 2003, **111**, 1877.
110. H. P. Ciolino and G. C. Yeh, *Br. J. Cancer*, 1999, **79**, 1340.
111. L. Macpherson and J. Matthews, *Cancer Lett.*, 2010, **299**, 119.
112. I. Fukuda, I. Sakane, Y. Yabushita, R. Kodoi, S. Nishiumi, T. Kakuda, S. Sawamura, K. Kanazawa and H. Ashida, *J. Agric. Food Chem.*, 2004, **52**, 2499.
113. D. Ung and S. Nagar, *Drug Metab. Dispos.*, 2007, **35**, 740.
114. M. Athar, J. H. Back, X. Tang, K. H. Kim, L. Kopelovich, D. R. Bickers and A. L. Kim, *Toxicol. Appl. Pharmacol.*, 2007, **224**, 274.
115. P. M. Kris-Etherton, K. D. Hecker, A. Bonanome, S. M. Coval, A. E. Binkoski, K. F. Hilpert, A. E. Griel and T. D. Etherton, *Am. J. Med.*, 2002, **113**(suppl. 9B), 71S.
116. P. Signorelli and R. Ghidoni, *J. Nutr. Biochem.*, 2005, **16**, 449.
117. S. Renaud and M. de Lorgeril, *Lancet*, 1992, **339**, 1523.
118. G. Lippi, M. Franchini, E. J. Favaloro and G. Targher, *Semin. Thromb. Hemost.*, 2010, **36**, 59.

119. H. P. Ciolino, P. J. Daschner and G. C. Yeh, *Cancer Res.*, 1998, **58**, 5707.
120. H. P. Ciolino and G. C. Yeh, *Adv. Exp. Med. Biol.*, 2001, **492**, 183.
121. B. D. Gehm, J. M. McAndrews, P. Y. Chien and J. L. Jameson, *Proc. Natl. Acad. Sci. USA*, 1997, **94**, 14138.
122. H. P. Ciolino and G. C. Yeh, *Mol. Pharmacol.*, 1999, **56**, 760.
123. E. Wenzel and V. Somoza, *Mol. Nutr. Food Res.*, 2005, **49**, 472.
124. R. F. Casper, M. Quesne, I. M. Rogers, T. Shirota, A. Jolivet, E. Milgrom and J. F. Savouret, *Mol. Pharmacol.*, 1999, **56**, 784.
125. A. H. Gilani, A. J. Shah, M. N. Ghayur and K. Majeed, *Life Sci.*, 2005, **76**, 3089.
126. P. Limtrakul, S. Lipigorngoson, O. Namwong, A. Apisariyakul and F. W. Dunn, *Cancer Lett.*, 1997, **116**, 197.
127. C. V. Rao, A. Rivenson, B. Simi and B. S. Reddy, *Cancer Res.*, 1995, **55**, 259.
128. H. Choi, Y. S. Chun, Y. J. Shin, S. K. Ye, M. S. Kim and J. W. Park, *Cancer Sci.*, 2008, **99**, 2518.
129. H. P. Ciolino, P. J. Daschner, T. T. Wang and G. C. Yeh, *Biochem. Pharmacol.*, 1998, **56**, 197.
130. A. L. Rinaldi, M. A. Morse, H. W. Fields, D. A. Rothas, P. Pei, K. A. Rodrigo, R. J. Renner and S. R. Mallery, *Cancer Res.*, 2002, **62**, 5451.
131. S. V. Singh, X. Hu, S. K. Srivastava, M. Singh, H. Xia, J. L. Orchard and H. A. Zaren, *Carcinogenesis*, 1998, **19**, 1357.
132. S. Gradelet, J. Leclerc, M. H. Siess and P. O. Astorg, *Xenobiotica*, 1996, **26**, 909.
133. P. Astorg, S. Gradelet, J. Leclerc, M. C. Canivenc and M. H. Siess, *Food Chem. Toxicol.*, 1994, **32**, 735.
134. P. Astorg, S. Gradelet, J. Leclerc and M. H. Siess, *Nutr. Cancer*, 1997, **27**, 245.
135. C. Jewell and N. M. O'Brien, *Br. J. Nutr.*, 1999, **81**, 235.
136. C. J. Gambone, J. M. Hutcheson, J. L. Gabriel, R. L. Beard, R. A. Chandraratna, K. J. Soprano and D. R. Soprano, *Mol. Pharmacol.*, 2002, **61**, 334.
137. Y. M. Yang, D. Y. Huang, G. F. Liu, J. C. Zhong, K. Du, Y. F. Li and X. H. Song, *Toxicol. Sci.*, 2005, **85**, 727.
138. K. Inouye, T. Mae, S. Kondo and H. Ohkawa, *Biochem. Biophys. Res. Commun.*, 1999, **262**, 565.
139. A. Jeuken, B. J. Keser, E. Khan, A. Brouwer, J. Koeman and M. S. Denison, *J. Agric. Food Chem.*, 2003, **51**, 5478.
140. I. A. Murray, J. L. Morales, C. A. Flaveny, B. C. Dinatale, C. Chiaro, K. Gowdahalli, S. Amin and G. H. Perdew, *Mol. Pharmacol.*, 2010, **77**, 247.
141. C. A. Flaveny, I. A. Murray, C. R. Chiaro and G. H. Perdew, *Mol. Pharmacol.*, 2009, **75**, 1412.

Small Model Organisms as Tools in Food Safety Research

MARIE TOHME,[a] JEAN-BAPTISTE FINI,[b] VINCENT LAUDET[a] AND BARBARA DEMENEIX[b]

[a] Institut de Génomique Fonctionnelle de Lyon, Ecole Normale Supérieure de Lyon, Université Lyon 1, CNRS, Université de Lyon, 46 allée d'Italie, 69364 Lyon Cedex 07, France; [b] CNRS UMR 7221, Département Régulations, Développement et Diversité Moléculaire, Muséum National d'Histoire Naturelle, 7 rue Cuvier, 7231 Paris Cedex 5, France

8.1 Introduction: the Physiological Relevance of Fish and Amphibian Embryo Small Model Organisms

The first six chapters of this book emphasize the current concerns about chemical contamination of the food supply and the potential risks and adverse effects of human exposure, particularly of unborn and newborn children, the growing child and adolescents. These different observations highlight the need for rapid screening methods that are both robust and reliable, but given the number of samples to be tested, not prohibitively expensive. Much international effort is currently directed to developing and validating methods that fulfil these criteria. However, the methods selected must also to take into account the 3Rs, *i.e.* the need to reduce, refine and replace animal testing by alternative methods. Most of such alternative methods are cellular- or molecular-based screening tests that focus on key aspects of a signalling process, for

Issues in Toxicology No. 11
Hormone-Disruptive Chemical Contaminants in Food
Edited by Ingemar Pongratz and Linda Vikström Bergander
© Royal Society of Chemistry 2012
Published by the Royal Society of Chemistry, www.rsc.org

instance interaction with the ligand binding domain of a receptor or activation of a given transcriptional or enzymatic process.

One notable advantage of most *in vitro* tests is their high-throughput capacity. Their two main disadvantages are, firstly, they are most commonly based on the use of single cell types or modelling of single receptor–ligand interactions and, secondly, in contrast to whole organisms or specialized cells such as hepatocytes, *in vitro* models will generally lack the full metabolic competence to convert, degrade and/or conjugate chemicals that *in vivo* models possess. Furthermore, as primary cell cultures are more difficult to obtain, most often transformed, immortalized cell lines are used for screening. The use of such cell lines reduces even more their physiological pertinence. Thus, if cell-specific assays are employed to inform regulatory decisions, then results from a spectrum of different cell-type or molecular-based assays need to be combined so as to obtain a realistic reconstruction of the manner that a given substance and/or its metabolites function in an organism. To overcome this limitation, an alternative test should possess not only the high-throughput advantage but also some of the high information content of *in vivo* tests. Effectively, the action of a given chemical in a complex multi-cellular organism will implicate multiple physiological factors, commonly grouped together under the umbrella term "ADMET" (absorption, distribution, metabolism, excretion and toxicology), many of these factors having tissue-specific characteristics.

Clearly, not all model organisms so far developed meet all of these requirements. Thus, regulators will in many instances have to consider data covering not only a maximum number of cell types or receptor-based assays but also possibly validated on a limited number of *in vivo* models. Hence, the higher the information content that can be achieved in the *in vitro* models, the greater their relevance to regulatory decisions will be. It is in this respect that fish and amphibian embryo models are proving so useful. As vertebrates they are, according to the signalling pathway under investigation, close or very close to mammals and humans. This similarity is particularly well exemplified by the case of three categories of nuclear receptors that include the much studied, common endocrine disruptor targets, the estrogen (ER), androgen (AR) and thyroid hormone (TR) receptors.[1] Taking the case of the TR, we note that the ligand triiodothyronine (T3) is exactly the same molecule in all vertebrates, from teleost fish (including zebrafish and medaka), to amphibians (including *Xenopus*) through to mammals (including mice and humans). Similarly, the ligand-binding domain (LBD) of the TR is highly conserved across vertebrates, as is its DNA-binding domain (DBD) and the most common DNA elements to which the TR binds in the genome (TRE, that are mainly variations on direct repeats of the PuGGTCA motif separated by 4 bp). Strong conservation is also found at the level of the thyroid gland across vertebrates, with the main components of thyroid hormone synthesis, *i.e.* iodination and organification, being common to all vertebrates from teleost through to mammals. Furthermore, many common distributor proteins (albumin, transthyretin) and activating/deactivating deiodinases are conserved even if some species-specific variations occur in some cases.[2] This exceedingly strong conservation of structural information and biochemical pathways makes the amphibian and teleost fish thyroid hormone signalling pathway entirely relevant for screening for potential thyroid hormone disruptors in both environmental and human health contexts.

A number of small model organisms (SMOs) have been used for screening purposes. Current favourites include the nematode or roundworm, *Caenorhabitis elegans*, the fruitfly, *Drosophila melanogaster*, and two vertebrates, the subjects of this review, the zebrafish, *Danio rerio*, and the anuran amphibian, *Xenopus laevis*, as well as the medaka, *Oryzias latipes*. Each of these models shares a number of key advantages: low stabulation costs, sizes compatible with large-scale screening programs in multiple-well plates and transparency of the embryo, allowing for easy detection of fluorescent protein expression in the living animal. Another major advantage is that each of these models is ideal for genetic modification, allowing the production of transgenics or mutants, *e.g.* for engineering specific reporter animals. The combination of each of these key features provides their overall advantage, that of permitting fluorescence-based high-throughput screening with a whole animal, *in vivo*.

However, despite these common features, each model has its advantages and disadvantages. Notably, *C. elegans* has a defined number of cells and is excellent for lineage studies, yet it possesses an extraordinary number of nuclear receptors that are duplicates of a unique gene, that encoding the orphan receptor HNF4. Furthermore, despite this large number of nuclear receptors, homologues for the main mammalian drug targets, including the ERs, ARs and TRs, are not present in the *C. elegans* genome. Given the presence of a strong cuticle in the adult and of the embryonic chorion, the fruit fly *D. melanogaster* is largely impermeable to chemical compounds and is thus not very easy to use as a chemical screening model. In addition, like *C. elegans* it lacks orthologues of the classical targets of endocrine disrupting chemicals (EDCs) in vertebrates, the ERs, the ARs and the TRs. In addition, these two models are invertebrates quite distant from vertebrates (specifically ecdysozoans, one of the two clades of the protostomes whereas vertebrates are part of the deuterostomes) and have been shown to be extremely derived compared to vertebrates and even to other protostomes. This phylogenetic distance implies that conclusions derived from these models cannot be easily extrapolated to either humans or to vertebrate wildlife. These limitations are not found in the free-living, vertebrate models, such as zebrafish, medaka and *Xenopus* that are thus becoming increasingly used as screening systems, but also as functional models that allow us to decipher the mechanisms of action of endocrine disruptors.

8.2   Small is Beautiful: the Size Advantage of Fish and Amphibian Larvae for Medium- to High-throughput Screening Methods

Over the last few years, much progress has been made in designing and producing transgenic small model vertebrate organisms for detecting chemicals in various substrates, including food extracts and environmental samples. The small size of the embryos means that they are easily placed in appropriate systems for medium- to high-throughput analysis either in 96-well plates or in flow-through systems.[3–5]

It is important to mention that adaptation of these aquatic vertebrate, fish and amphibians to medium or high throughput requires use of early embryos. If the early-stage embryos chosen are non-feeding stages (as is the case for *Xenopus laevis* up to stage 45, or approximately seven days post-fertilization), they enter the legislative category of *in vitro* models. Other advantages of each model will be described below.

The development of flow-through systems has the distinct advantage that the larvae do not have to be handled for placement in a multi-well system. The elimination of this step has two major benefits. First of all it reduces costs, as it eliminates the labour-intensive step of placing individual larvae in the wells. Second, it has the major advantage of avoiding any unnecessary stress to the larvae that could be induced by the transfer to the wells.[4]

8.3 Metabolic Capacity of Fish and Amphibian Larvae

The main advantage of using *in vitro* stages of small model organisms such as *Xenopus* and zebrafish is that they combine the high-content aspects of whole organism approaches with the high-throughput aspects of robotizable methods. Notably, screening in 96-well plates is feasible for early developmental stages of both organisms, thereby accelerating the number of chemicals that can be analyzed. However, it is clear that one of the principal advantages of *in vivo*, high-content assays is the capacity to faithfully represent the metabolic pathways. Recent work has shown that *Xenopus* tadpoles have the capacity to metabolize the well-characterized endocrine disrupting chemicals bisphenol A (BPA) and tetrabromobisphenol A (TBBPA), in a manner similar to that found in mammals. These results indicate that *Xenopus* shares similar phase II metabolic pathways (glucuronidation, sulfation) to mammals.[6,7]

Recently, we addressed the biotransformation of TBBPA by *Xenopus laevis* tadpoles. The kinetics of TBBPA uptake from and release into aquarium water were evaluated using radiolabelled TBBPA. The HPLC main metabolites of TBBPA were separated, characterized and analyzed. The profiles showed varying proportions of glucuronide-TBBPA, TBBPA-glucuronide-sulfate, sulfate-TBBPA and bisulfate-TBBPA. These metabolites are similar to those seen in mammalian systems, underlining that the pathways used to metabolize, conjugate and facilitate excretion are parallel in these two vertebrate classes.[7]

The metabolic ability of zebrafish or medaka embryos with respect to endocrine disruptors is beginning to be studied and it seems that the same principles acting in mammals are also present in fish. The metabolism of steroid hormones, for example, is strongly conserved between teleost fishes and mammals.[8] The ability of several P450 enzymes to be regulated by xenobiotics has been studied in several fish models. Of course, some species-specific effects have been reported, but the main principles of these regulations are conserved. Finally, as regards fish nuclear receptor signalling, both the receptors themselves and the main targets of pollutants and endocrine disruptors have been extensively studied. As for the *Xenopus* model, structures and targets are similar to those in mammals, although some differences in gene number occur.[9]

One particularly relevant example of metabolic effects is that of bisphenol. We have recently described a new phenotypic effect elicited by exposure to bisphenol A during zebrafish embryogenesis.[10] We observed that bisphenol A is able to induce abnormalities of the otic vesicle, the future inner ear of the adult fish. In both zebrafish- and *Xenopus*-treated embryos we observed abnormal aggregates of the otoliths, the calcium carbonate structures that allow the transmission of vibration and acceleration forces required for hearing and balance. This effect was elicited by bisphenol A but not by bisphenol F, a derivative in which the two methyl groups of the central carbon are missing. Interestingly, using ^{14}C-labelled compounds we observed that bisphenol A readily entered and accumulated in the zebrafish and can be found unmetabolized 24 hours later, suggesting that no major degradation pathway had occurred. This lower metabolism parallels that seen for BPA in *Xenopus* and can be likened to the situation in human samples,[4,6] where BPA, but not its metabolites, are largely present in urine samples. In contrast, the situation is different for bisphenol F. Indeed, with this compound, if we also observed an accumulation within the zebrafish embryo, we detected a degradation product suggesting that BPF, but not BPA, was metabolized into a non-identified derivative within the zebrafish embryo. This may explain, at least in part, the inability of BPF to induce the otolith aggregates that are observed after BPA treatment. This example illustrates the importance of these metabolic results for explaining the divergent effects of related products, a feature commonly seen in toxicology testing.

8.4   Experimental Approaches

Both fish and amphibian models can be used to dissect the role of specific genes as mediators of the effects of endocrine disruptor substances. Such approaches can employ specific systems such as morpholinos directed against target mRNAs to knock-down gene expression, pharmacological products (*e.g.* agonists or antagonists of known signalling pathways), or mutants and reporter models. This wide panoply of techniques allows the researcher to address functional questions with *in vivo* experiments.

8.4.1   Gene Knock-down

In both fish models and *Xenopus* it is possible to perform gene knock-down, using specific reagents known as morpholinos, *i.e.* modified, stable antisense oligonucleotides that can block specifically either translation of an mRNA or its splicing. However, the RNA interference (RNAi) system, often used in other organisms, has yet to be demonstrated as functional in zebrafish or medaka. In *Xenopus*, some proofs of principle have been shown but this approach is still only used rarely. As zebrafish or *Xenopus* eggs are relatively big and spawned in large numbers, it is relatively easy to inject these eggs with morpholinos targeting specific genes to study the developmental consequences of this inactivation. This is a quite effective and easy system to test the function of a given gene in an *in vivo* context and this is even amenable to medium-throughput screening. One further interest is that it is possible to test the effects of morpholinos targeting two (or even three) different genes, permitting one to decipher the functions of closely related genes, such as those produced by the whole genome duplications that arose early during actinopterygian fish evolution. Nevertheless, the use of morpholinos is not without possible problems. As some unspecific toxicity may occur, several specific controls should be carried out. For example, the effect of two different morpholinos targeting the same gene but at different places (*e.g.* one translation blocker and one splice blocker) should be compared. Ideally a decrease in the expression of the targeted mRNA (splice blocker) or protein (translation blockers) should be observed and a rescue of the effect by the injection of an mRNA coding for the targeted protein (of course engineered as to avoid the blocking by the morpholino) should be observed. Another problem is that given that the morpholinos are injected in one-cell-stage eggs, they can only produce effects during the first 72 hpf (hours post-fertilization) of development, thus limiting the effect to the early life of fish or frog larvae. Nevertheless, as this period covers the whole embryonic period (hatching starts at 48 hpf), the scope of possible relevant information that can be gathered with morpholinos is quite large. Finally, the biggest drawback of morpholinos is that even with all the necessary controls these reagents only decrease the expression of the gene targeted but do not produce real loss of function; that is, with a morpholino approach one can have gene knock-down but not gene knock-out and this may render some interpretations quite difficult.

Interestingly, a new system, the zinc-finger nuclease (ZFN), has recently been introduced into the zebrafish tool kit. This system overcomes many of the limitations of the morpholinos. ZFNs, which are a chimeric fusion between a Cys2-His2 zinc-finger protein (ZFP) and the cleavage domain of FokI endonuclease, were originally developed for genomic manipulation in plants, invertebrates and cell lines. The DNA-binding specificity is supplied by the ZFP, which can be engineered to recognize a wide variety of target sequences, and the cleavage activity is provided by the nuclease domain. The nuclease domain requires dimerization for activity, and consequently two ZFNs must assemble on DNA with the appropriate geometry for efficient DNA cleavage. ZFN-mediated cleavage can lead to mutations when double-stranded breaks are repaired by nonhomologous end joining (NHEJ), an error-prone repair pathway. Thus, ZFNs allow directed mutagenesis of a desired target sequence in a complex genome and this technology has been now been applied in zebrafish, with very promising results.[11,12] This will allow us to have null mutants for the main endocrine disruptor targets such as ERs, TRs or ARs and to test in a genetically clear context the role of each receptor. Interestingly, combined with morpholino inhibition, this will give rise to an extremely efficient system, allowing analysis of functionally complex situations in which several genes are acting.

8.4.2 Pharmacological Testing

As the main features of endocrine systems are conserved across vertebrates, it is possible to use the knowledge obtained in mammals to test for the implication of a given pathway in an observed effect and *vice versa*. For example, as mentioned earlier, the TRs or the ERs are well conserved between teleost fishes and mammals. Thus, treating the teleost or *Xenopus* SMOs with known agonists or antagonists of these receptors allows one to test if those receptors are implicated in a given response.

This approach was used in the work on the otic vesicle phenotype observed in zebrafish after BPA treatment.[10] The effect observed after BPA treatment (at doses >5 M) cannot be observed using either 17β-estradiol (E2), an ER agonist, or ICI182780, an ER antagonist. This finding suggests that simply activating or blocking the ERs is not sufficient to elicit the effect. Furthermore co-treatments with BPA and either E2 or ICI do not block the effect of BPA, further suggesting that the estrogen receptors are not directly involved in this phenotype. Similar results have been obtained with thyroid hormones, allowing us to exclude the TRs in this effect.

However, this type of analysis should be performed with caution. Indeed, the selectivity and specificity of the compounds used have to be first evaluated *in vitro* on the receptors targeted if one wants to obtain truly relevant results. It is not because a compound is selective on human receptors that this selectivity is strictly conserved in other species. For example, we have observed that genistein is an ERα-selective compound in mammals but not in zebrafish, probably because of a small number of mutations present in the zebrafish ERs.[13] However, if these cautions are fully taken into account, then a pharmacological approach can be rapid and efficient to delineate the possible targets implicated in any phenotypic effect and this could be much more rapid than a genetic strategy. Indeed, when both pharmacological approaches and gene knockdown strategy converge towards the same results, this renders very high the quality of the inference.

8.5 Specific Models for Screening

8.5.1 THbZIP-GFP *Xenopus* for Thyroid Hormone Signalling

Given the absolutely essential role of the thyroid hormone (TH) in brain development of the developing fetus and child, screening for thyroid hormone disruptive mixtures in foodstuffs,[14] even at low doses, is critical. When searching for an alternative model to test for thyroid hormone disruption it was logical to consider the potential of amphibian models and particularly the

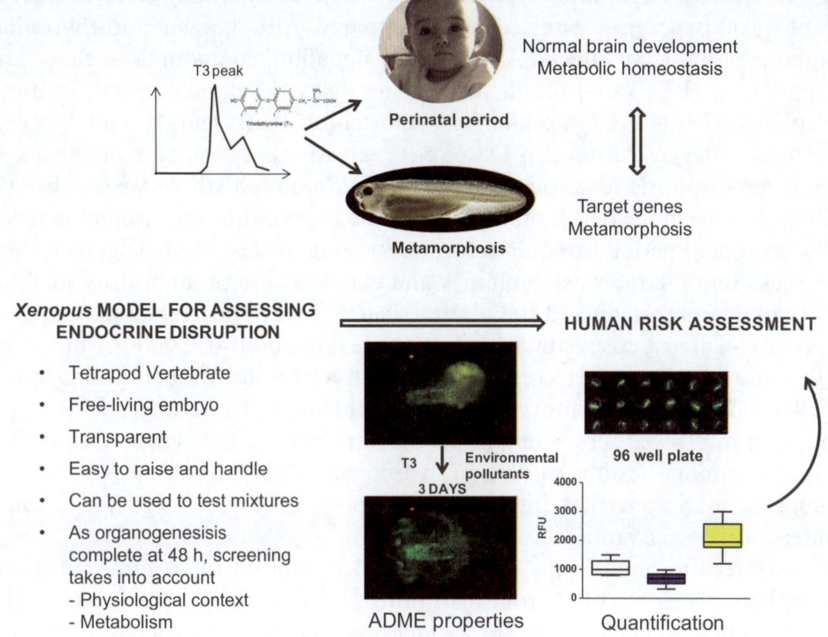

Figure 8.1 The *Xenopus* embryonic thyroid-disruption assay (XETA) and its pertinence to human risk assessment. The principle of the XETA test is that the embryonic *Xenopus* is competent to respond to TH and therefore can be used to detect disruption of TH signalling. TH is exactly the same molecule in all vertebrates and many of the main elements of thyroid signalling are shared between vertebrates (notably, receptors, metabolism and many target genes). Thus, screening in *Xenopus* has high relevance for human toxicology. Furthermore, both *Xenopus* and humans have key developmental periods critically dependent on TH: the perinatal period in humans and metamorphosis in amphibians.

process of amphibian metamorphosis. This reasoning is based on two important and well-established facts. First, it should be emphasized that TH is exactly the same molecule in all vertebrates, from fish, through amphibians and reptiles, to birds and mammals. Similarly, the molecular machinery for producing, activating and transducing the TH signal is highly conserved throughout vertebrates. The second major idea is the fact that TH orchestrates the metamorphic programme and the same hormone is vital to the perinatal period in mammals, including humans (Figure 8.1). Indeed, lack of TH during the first years of life in humans causes irreversible brain damage, or cretinism.

The fact that TH is essential for a tadpole to change into a frog during metamorphosis was the basis for the development and optimization of the OECD amphibian metamorphosis test (OECD 231). This test uses NF51 premetamorphic *Xenopus laevis* tadpoles placed for 21 days in the presence of the molecule to test. After this 3-week paradigm, either morphological issues (developmental stage, inter-eyes distance, whole body length, limb length) or histological issues are assessed and compared with known anti-thyroidian compounds inducing effects. Anti-thyroid signalling compounds such as propylthiouracil (PTU) or sodium perchlorate delay development and, in metamorphosis, T3-induced mechanisms (shortening body length, limb growth, shortened inter-eye distance). Despite its sensitivity, two major drawback of this metamorphosis assay are, first, the time required (three weeks) for the tadpoles to reach the developmental stage used for testing and, second, a three-week treatment period is required for the running the test itself. Together, these six weeks imply large waste volumes and extended use of laboratory facilities for amphibian breeding and stabulation, factors that in turn make the test quite expensive. Current cost estimates place the test at about €30 000 per molecule.

In contrast, the more recently developed XETA uses much younger and smaller tadpoles at the embryonic/early larval stage. These stages are compatible with multi-well screening and as they are non-feeding stages they do not fall under animal testing legislation in Europe.[3] This fluorescence-based physiological test is one that integrates all aspects of thyroid signalling, which confers a major advantage.

This screening model was developed and optimized for routine, medium-throughput screening of thyroid disrupting chemicals using the ThbZIP-GFP *Xenopus* line.[3,15] This transgenic *Xenopus laevis* line bears genetic constructs integrating thyroid hormone responsive elements (TREs) cloned upstream to the green fluorescent protein coding sequence. OECD criteria of disruption of the thyroid axis were used to assess signals derived from fluorescent embryos placed in 96-well plates. Benchmarking has focused on defining the specificity and sensitivity of the method. To determine specificity, a series of food additives and pharmaceutical drugs with no known activity on the thyroid axis were selected. To determine the sensitivity of the test and its capacity to detect the disrupting potential of chemicals, substances known to modulate thyroid signalling at different levels were selected, based on results obtained in the OECD amphibian metamorphosis assay (OECD 231). The protocol developed using embryonic, nonfeeding stages of embryos for three days, integrates reliable parameters and provides reproducible results that are coherent with those obtained in the metamorphosis assay. The methodology is robust and can now be used for cross-laboratory validation studies.

8.5.1.1 Substances of Particular Concern for Thyroid Disruption

8.5.1.1.1 Perchlorate. Many pollutants have the potential to interfere with thyroid hormones.[16,17] One major current concern is the presence of perchlorate and other thyroid disrupting factors in foodstuffs, notably milk, largely consumed by babies and children. Perchlorate is of particular concern as this thyroid disrupting substance has been shown to accumulate in human and cows' milk.[18] Perchlorates are salts derived from perchloric acid ($HClO_4$) and much used in the explosives industry, especially in pyrotechnics and as a component of solid rocket fuel. The salts are iodine-competitive inhibitors which act at the sodium/iodine symporter level, localized at the basal layer of the thyroid, salivary and mammalian glands and intestine.[19] Most perchlorate salts are soluble in water and can be found in significant levels in tap water, particularly around rocket test sites. That the action of perchlorates can be detected by a transgenic *Xenopus* line was reported in 2007.

8.5.1.1.2 Polychlorobiphenyls. Polychlorobiphenyls (PCBs) represent a family of 209 congeners, used extensively between 1930 and 1950 in pesticides, paints and flame retardants. Owing to their lipophilic properties, PCBs are part of the persistent organic pollutant (POP) family that bioaccumulate in organisms and sediments. For this reason they were banned from plastics, adhesives and inks and for flame retarding uses in 1976 in the USA and three years later in Europe.[20] PCBs and mostly their hydroxylated metabolites have a strong structural homology with the thyroxine T4. PCBs exposure correlates with low TH levels, particularly free T4.[21–23] Monkeys orally exposed to PCBs for 18–23 weeks show a significant diminution in T4, T3, free T4 and a raise in TSH levels.[24] Interestingly, these variations are observed in hypothyroidian patients. *In utero* exposure could be a problem also. For chicken eggs, PCB exposure causes a significant delay in hatching and alters the timing of the T3 peak, which occurs in chickens before hatching.[25,26] It is noteworthy that a T3 peak occurs in all vertebrates at key developmental stages that often imply major lifestyle switches. In mammals, including humans, it occurs at birth and it occurs at metamorphosis in amphibians (Figure 8.1). In human IQ, defects have been reported for children exposed during their mother's pregnancy and PCBs have been shown to delay metamorphosis in *Xenopus laevis*.[27,28] Other effects on the reproductive system have been attributed to PCB exposure in fish and amphibian models; however, whether the effects of PCB thyroid signalling can be detected using fluorescent, transgenic lines has yet to be assessed.

8.5.1.1.3 Bisphenol A. Bisphenol A (BPA) was first synthesized in the 1930s as an estrogenic replacement.[29] However, owing to its weak estrogenicity it was replaced by diethylstilbestrol (DES), with dramatic consequences. Indeed, DES was extensively prescribed to reduce the risk of miscarriage between 1950 and 1970 in the USA and Europe, with the unexpected and unfortunate appearance of rare cancers in young girls and gonadal development problems for both girls and boys born to treated women.

In the 1970s, BPA was re-used for its polymerization properties. The resulting polymers are soft but resistant and transparent. The first pharmacological studies indicated that BPA was non-accumulative.

Thus BPA rapidly became one of the most produced chemicals, with a production rate of over 3.8 million tonnes per year (1.15 in Europe).[30] The acceptable daily intake (ADI) is fixed at 0.05 mg kg^{-1} in the USA and Europe.

Its use in baby feeding bottles has been controversial and banned in California and Canada in 2008 and in France in 2010. The literature is replete with papers showing BPA to have similar effects to DES.[31] *In utero* exposure to low doses of BPA (2.5–250 µg kg^{-1}) alters normal development of mammary glands in young mice.[32] Even if effects on the thyroid are less studied than reproductive effects, anti-thyroidal actions of BPA have been reported. Exposed rats show increased thyroid weights.[33] Further, positive correlation between BPA exposure and the activity of detoxification enzymes has been demonstrated.[34] This conjugation pathway is also used for TH metabolism,[35] indicating a route for BPA disruption of thyroid signalling. Another level of disruption could be at the level of the thyroid receptor (TR), as Moriyama *et al.* showed that BPA could bind weakly to TRs.[36] As regards the detection of anti-thyroid activity using vertebrate SMOs, Fini *et al.* showed in 2007 that BPA has anti-thyroidian effects in *Xenopus laevis* TH reporter animals.[3] These results were confirmed more recently by authors using a more classical *in vivo* metamorphic assay.[37]

8.5.1.1.4 Tetrabromobisphenol A. Since the banning of PCBs, the most-used brominated flame retardant in the world has become tetrabromobisphenol A (TBBPA), with an annual production over 210 000 tonnes.[38] Early pharmacokinetic studies showed it to be rapidly excreted without undergoing metabolism. More recent studies confirmed that TBBPA is rapidly excreted both in rodents and in humans, but showed that BPA undergoes metabolism in both groups.[39,40]

Despite the close structural similarity of TBBPA to TH (the four brominated atoms are placed like the four iodine atoms in thyroxine), studies on TBBPA effects on thyroid signalling are quite recent. *In vitro* studies showed competition with TH at the receptor level; however, TBBPA has an affinity three orders of magnitude less.[41] Meerts *et al.* showed TBBPA to bind to transthyretin (a serum protein that acts as distributor for TH) with a 10-fold greater affinity than T4.[42] Other *in vivo* studies in *Xenopus* showed TBBPA to inhibit TH-induced tail regression and to delay TH-dependent metamorphosis.[43,44] These anti-thyroid signalling effects of TBBPA were also detected when using transgenic larval *Xenopus* in which GFP expression is driven by a TH responsive gene promoter (THbZIP).[3]

Endocrine disruption can also be the result of production of metabolites rather than being caused by the parent compound. This possibility emphasizes the need for screening systems that integrate metabolism. We recently showed that BPA was partially metabolized by *Xenopus laevis* and that TBBPA was extensively metabolized by the same animals.[4,6,7]

8.5.1.1.5 Perfluoro Compounds. Perflurooctanesulfonate (PFOS) and per-fluorooctanoic acid (PFOA) are compounds used in anti-adhesive frying pans and saucepans. They are found to be accumulative in humans and are inversely correlated with TH levels. Most of the studies have dealt with human levels.[45] We have been able to detect significant anti-thyroidian effects of

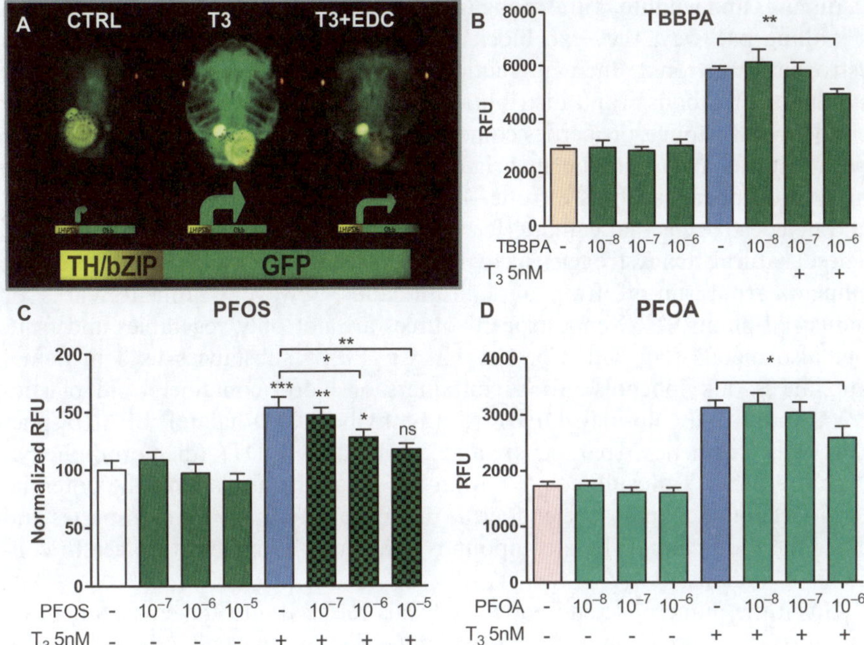

Figure 8.2 Stage 45 TH reporter tadpoles can be used to detect different categories of thyroid disruptors. (A) Principal of the *Xenopus* transcriptional assay. Control (CTRL) animals show basal GFP expression (green fluorescence) driven by the THbZIP promoter; T3 induces up-regulation of GFP after 72 h T3 (5 nM); T3 + EDC (here 1 µM TBBPA) induces an intermediary level of activation, revealing an anti-thyroid hormone effect of TBBPA. (B) Transcriptional responses in tadpoles treated with various concentrations of TBBPA using quantification of fluorescence. (C) Transcriptional responses in tadpoles treated with various concentrations of perfluorooctanesulfonate (PFOS). (D) Transcriptional responses in tadpoles treated with perfluorooctanoic acid (PFOA). THbZIP-eGFP transcription was measured using an Ultra Evolution multiple-well plate reader. Values shown are means [SE (*n*) 20 per group]. Each experiment was done at least three times, providing the same result. Statistical differences were assessed using the Mann-Whitney non-parametric test; *, $p < 0.05$, **, $p < 0.01$, ***, $p < 0.001$.

both PFOS and PFOA using our THbZIP-GFP *Xenopus* system (Figures 8.2C and 8.2D; unpublished data).

8.5.2 Zebrafish as Models to Detect Disruption of Estrogen Signalling

Many pollutants released into the environment have the potential to disrupt the normal function of the endocrine system and cause reproductive impairments in humans and wildlife.[46,47] Among these compounds, many target the estrogen signalling pathway: they can block or mimic the activity of the endogenous estrogens, antagonize their interaction with estrogen receptors or affect their synthesis, metabolism and distribution in the organism. Chemicals that show endocrine disrupting properties come from a range of chemical groups that are both natural and synthetic in origin. Humans, livestock and plants excrete natural chemicals such as estrone, 17β-estradiol, 16α-hydroxyestrone, phyto- and mycoestrogens that can interfere with the estrogen signalling pathway.[48,49] These natural xenoestrogens have been detected in surface water receiving effluents from run-off from agriculture lands, sewage treatment works or industrial plants.[50,51] Xenoestrogen sources are not only vegetables and fruit, but also metals (Hg, Cd, Co, Cu, Ni, Cr, Pb),[52] substances used in dental appliances (alkylphenols), food containers or blood containers and plastics [PVC (polyvinyl chloride), DEHP (di-(2-ethylhexyl) phthalate), BPA (bisphenol A)], cosmetics (parabens) and pesticides [DDT (dichlorodiphenyl-trichlorethane), endosulfane].[53–55] Many of these chemicals have the potential to bioaccumulate and cause endocrine disruptions to many animal species and enter the food chain. These compounds can thus ultimately cause health concerns in humans.

Given the numerous xenobiotic chemicals released into the environment by human activity that have the potential to disrupt the endocrine system in wildlife and humans, current questions include what mechanisms are involved, the effective concentrations and whether disruption will lead to ecologically relevant effects.[56] With this in mind, novel test systems have been developed exploiting aquatic species to determine endocrine activities of chemicals, and to study the mechanisms of action of known endocrine disrupters. Zebrafish (*Danio rerio*) and medaka (*Oryzias latipes*) fish offer technical and practical advantages for studying principal biological processes, effects and mechanisms. Such models have particularly focused on estrogen signalling. Below, we describe the main tools available in terms of transgenic reporter lines that allow one to (i) detect the presence of an estrogenic activity elicited by a given compound or by a mixture of chemicals and (ii) better understand the mechanisms of action of certain chemicals.

Legler *et al.* developed a stably transfected zebrafish line containing a luciferase gene under control of a minimal promoter linked to an estrogen response element that can be recognized by the endogenous estrogen receptors.[57] Short-term exposure of *ERE-Luc* transgenic zebrafish to estrogenic compounds leads to *in vivo* ER transactivation followed by luciferase gene activation and increased luciferase activity that can be measured in tissue lysates by measuring light activity following addition of the enzyme substrate luciferin. When juvenile transgenic zebrafish were exposed to low concentrations of estradiol (E2), estrone, estrogens diethystilbestrol (DES) or 17α-ethynylestradiol (EE2), the period of gonad differentiation was shown to be the most responsive early life stage. In the case of E2, this resulted in 100-fold induction of the luciferase at 1000 nM compared to control, after 96 h of exposure of 35 days for juvenile fish (the stage where they undergo sexual differentiation). More importantly, *o,p'*-DDT, the principle DDT isomer, also induced luciferase activity in *ERE-Luc* transgenic fish. DDT, an organochlorine pesticide, was one of the most largely produced and widely dispersed in the environment.

The advantage of this novel *in vivo* test system is that it allows one to measure rapidly the effects of estrogenic chemicals on critical life stages and sensitive target organs in fish. However, a major drawback of this assay is that the reporter luciferase is not directly visible in live fish, and lysates of pooled embryos or tissues from adults have to be prepared.

Another zebrafish transgenic line that has been developed for rapid detection of environmental estrogens is *ere-zvtg1: gfp*, in which the green fluorescence protein (GFP) reporter gene is under control of the zebrafish vitellogenin1 (zvtg1) promoter inserted downstream of a 39 bp DNA sequence containing the consensus estrogen response element (ERE).[58] Vitellogenin is a protein accumulated by females in the eggs and which is expressed at high levels in the liver under estrogen control. This gene is naturally estrogen-responsive, but the authors that have constructed this reporter line have improved its estrogen sensitivity by adding a consensus ERE. When exposed to very low concentrations of EE2, E2, estriol, DES, progesterone or 17-hydroxy steroid, an enhanced expression of GFP is established in the *ere-zvtg1: gfp* transgenic zebrafish line, showing very bright fluorescence in the liver in both larvae and adult fish. For example, embryos exposed at the one- and two-cell stages to concentrations from 0.1 to 100 ng L^{-1} EE2 showed GFP expression in the liver. Interestingly, larvae exposed to cadmium chloride ($CdCl_2$) or BPA also showed observable GFP expression in the liver. Another fish transgenic line similar to *ere-zvtg1: gfp* has been developed. This is the *mvtg1: gfp* transgenic medaka line in which the gfp reporter gene was placed under the control of the medaka vitellogenin1 (mvtg1) gene promoter.[59] Similar to the *ere-zvtg1: gfp* transgenic zebrafish line, the medaka *mvtg1: gfp* transgenic line express GFP exclusively in the liver in juvenile and mature fish, when exposed to E2 (1 µg L^{-1}), EE2 (1 µg L^{-1}), DES (5 µg L^{-1}), estriol (10 µg L^{-1}) or BPA (5000 µg L^{-1}). Both these transgenic lines appear to induce GFP expression in the liver. However, the medaka transgenic line seems to be less sensitive than the zebrafish transgenic line for detecting estrogens. Thus, (i) the exposure time to chemicals to induce GFP expression is longer at the same concentrations of EE2 (1 µg L^{-1}): in medaka *mvtg1: gfp*, 21 days treatment were necessary, whereas for the zebrafish *ere-zvtg1: gfp* transgenic fish were treated for only 4 days to obtain a strong GFP induction. (ii) At low concentrations, zebrafish *ere-zvtg1: gfp* transgenic fish do not need dissection; GFP fluorescence can be visible through the live fish abdomen wall when exposed to EE2, which is not the case with the medaka *mvtg1: gfp* transgenic fish.

Two other transgenic medaka lines with a GFP-reporter gene, driven either by choriogenin H gene regulatory elements or by choriogenin L gene regulatory elements,[60,61] have been developed. Choriogenins (chg-H, chg-L) are precursor proteins of the egg envelope of medaka and synthesized in the spawning female liver in response to estrogen. When exposed for 24 h to E2 (0.63 nM), EE2 or estrone (14.8 nM), transgenic medaka larvae developed GFP expression in the liver.

Another transgenic zebrafish line (*hsp70-4/eGFP*) has been developed as an *in vivo* tool, to measure cadmium, arsenic and mercury pollution.[62] Heat shock proteins (HSPs) are a family of highly conserved molecular chaperones that serve in the proper folding, transport and degradation of cellular proteins. Different cytotoxic agents, including heavy metals such as mercury, cadmium, *etc.*, have been shown to induce hsp70 expression. In normal conditions, heat shock transcription factors (HSFs) are in an inactive form and are present as hetero-oligomers with hsp70. When cells are exposed to stresses, proteins unfold, and nonnative proteins present a higher affinity for the hsp70 than the native ones. Under stress conditions, hsp70 is released from HSF in the cytoplasm, allowing HSF to assemble DNA binding homotrimers that migrate to the nucleus and then bind the heat shock element (HSE) in the promoter region of the hsp genes and enhance these genes transcription. When cells are exposed to metals (*e.g.* cadmium), the metals interact with SH-group proteins and alter their conformation due to an oxidative action, thus stressing the cells and thereby inducing hsp synthesis. Based on these studies, researchers have developed transgenic lines using the promoter of the HSP70 gene linked to GFP. Transgenic zebrafish juveniles exposed to cadmium showed an accurate and reproducible level of GFP expression due to the hsp70-4 promoter activation. Cells in the gills and skin showed detectable fluorescence in live larvae at 24 h after 3-h cadmium exposure at 0.2 µM, allowing maximal GFP accumulation. The observed GFP expression shows a dose-dependent response from 0.2 µM to 125 µM. At the highest cadmium concentration, GFP expression was observed in the olfactory organ, gills, skin, liver and pronephros of the fish

These data illustrate that fish eggs, juvenile or adult fish, designed as biomarkers, are promising new tools in the field of environmental contaminant monitoring and risk assessment of new and existing chemicals.

8.6   Conclusions and Perspectives

Clearly the numerous advantages of these vertebrate, fish and amphibian, small models are leading to their use in multiple applications, not only for detecting contaminants in the environment and in food samples, but also in pharmaceutical research. These SMOs are based on physiological effects of chemicals on target organs and the whole body. They are excellent tools offering the biotechnology industry a new, cheaper and faster alternative for the production of recombinant proteins that require complex patterns as post-translational processing. No doubt one key factor driving their popularity is the versatility of their use, with infinite potential combinations of fluorescent markers and genetic response elements to be exploited. As to the future, one expects the proximity of a number of their signalling pathways to those in humans to lead to the design of models of human disease. Design and production of novel fluorescent vertebrate SMOs that can be screened in multi-wells could accelerate and redirect key areas of drug research and development.

Acknowledgements

Work from our laboratories is funded by the EU (Cascade, Crescendo, SME-Receptor) as well as by the Agence Nationale de la Recherche (Kismet) and PNRPE.

References

1. H. Gronemeyer, J. A. Gustafsson and V. Laudet, *Nat. Rev. Drug Discov.*, 2004, **3**, 950–964.
2. M. Paris and V. Laudet, *Genesis*, 2008, **46**, 657–672.
3. J. B. Fini, S. Le Mevel, N. Turque, K. Palmier, D. Zalko, J. P. Cravedi and B. A. Demeneix, *Environ. Sci. Technol.*, 2007, **41**, 5908–5914.
4. J. B. Fini, S. Pallud-Mothre, S. Le Mevel, K. Palmier, C. M. Havens, M. Le Brun, V. Mataix, G. F. Lemkine, B. A. Demeneix, N. Turque and P. E. Johnson, *Environ. Sci. Technol.*, 2009, **43**, 8895–8900.
5. R. T. Peterson, B. A. Link, J. E. Dowling and S. L. Schreiber, *Proc. Natl. Acad. Sci. USA*, 2000, **97**, 12965–12969.
6. J. B. Fini, L. Dolo, J. P. Cravedi, B. Demeneix and D. Zalko, *Ann. N.Y. Acad. Sci.*, 2009, **1163**, 394–397.
7. J. B. Fini, L. Debrauwer, A. Hillenweck, S. Le Mével, S. Chevolleau, A. Boulahtouf, K. Palmier, P. Balaguer, J.-P. Cravedi and B. Demeneix, *Toxicol. Sci.*, 2011, submitted.
8. R. Thibaut and C. Porte, *J. Steroid Biochem. Mol. Biol.*, 2004, **92**, 485–494.
9. S. Bertrand, B. Thisse, R. Tavares, L. Sachs, A. Chaumot, P. L. Bardet, H. Escriva, M. Duffraisse, O. Marchand, R. Safi, C. Thisse and V. Laudet, *PLoS Genet.*, 2007, **3**, e188.
10. Y. Gibert, S. Sassi-Messai, J. B. Fini, L. Bernard, D. Zalko, J. P. Cravedi, P. Balaguer, M. Andersson-Lendahl, B. Demeneix and V. Laudet, *BMC Dev. Biol.*, 2011, **11**, 4.
11. Y. Doyon, J. M. McCammon, J. C. Miller, F. Faraji, C. Ngo, G. E. Katibah, R. Amora, T. D. Hocking, L. Zhang, E. J. Rebar, P. D. Gregory, F. D. Urnov and S. L. Amacher, *Nat. Biotechnol.*, 2008, **26**, 702–708.
12. X. Meng, M. B. Noyes, L. J. Zhu, N. D. Lawson and S. A. Wolfe, *Nat. Biotechnol.*, 2008, **26**, 695–701.
13. S. Sassi-Messai, Y. Gibert, L. Bernard, S. Nishio, K. F. Ferri Lagneau, J. Molina, M. Andersson-Lendahl, G. Benoit, P. Balaguer and V. Laudet, *PLoS One*, 2009, **4**, e4935.
14. J. Bernal, *Nat. Clin. Pract. Endocrinol. Metab.*, 2007, **3**, 249–259.
15. N. Turque, K. Palmier, S. Le Mevel, C. Alliot and B. A. Demeneix, *Environ. Health Perspect.*, 2005, **113**, 1588–1593.
16. F. Brucker-Davis, *Thyroid*, 1998, **8**, 827–856.
17. E. N. Pearce and L. E. Braverman, *Best Pract. Res. Clin. Endocrinol. Metab.*, 2009, **23**, 801–813.
18. L. E. Braverman, *Thyroid*, 2007, **17**, 819–822.
19. J. Wolff, *Pharmacol. Rev.*, 1998, **50**, 89–105.

20. J. L. Lincer and D. B. Peakall, *Nature*, 1970, **228**, 783–784.
21. S. A. van der Plas, I. Lutkeschipholt, B. Spenkelink and A. Brouwer, *Toxicol. Sci.*, 2001, **59**, 92–100.
22. S. Hallgren, T. Sinjari, H. Hakansson and P. O. Darnerud, *Arch. Toxicol.*, 2001, **75**, 200–208.
23. S. Hallgren and P. O. Darnerud, *Toxicology*, 2002, **177**, 227–243.
24. K. J. van den Berg, C. Zurcher and A. Brouwer, *Toxicol. Lett.*, 1988, **41**, 77–86.
25. V. Beck, S. A. Roelens and V. M. Darras, *Gen. Comp. Endocrinol.*, 2006, **148**, 327–335.
26. S. A. Roelens, V. Beck, J. Maervoet, G. Aerts, G. E. Reyns, P. Schepens and V. M. Darras, *Gen. Comp. Endocrinol.*, 2005, **143**, 1–9.
27. J. L. Jacobson and S. W. Jacobson, *New Engl. J. Med.*, 1996, **335**, 783–789.
28. E. A. Lehigh Shirey, A. Jelaso Langerveld, D. Mihalko and C. F. Ide, *Environ. Res.*, 2006, **102**, 205–214.
29. E. C. Dodds, L. Golberg, W. Lawson and R. Robinson, *Nature*, 1938, **141**, 247–248.
30. E. Burridge, *Eur. Chem. News*, 2003, 14–20 (Europe Plastics 2006 http://www.plasticseurope.org/).
31. G. Schonfelder, B. Flick, E. Mayr, C. Talsness, M. Paul and I. Chahoud, *Neoplasia*, 2002, **4**, 98–102.
32. M. Durando, L. Kass, J. Piva, C. Sonnenschein, A. M. Soto, E. H. Luque and M. Munoz-de-Toro, *Environ. Health Perspect.*, 2007, **115**, 80–86.
33. B. L. Tan, N. M. Kassim and M. A. Mohd, *Toxicol. Lett.*, 2003, **143**, 261–270.
34. M. Boas, U. Feldt-Rasmussen, N. E. Skakkebaek and K. M. Main, *Eur. J. Endocrinol.*, 2006, **154**, 599–611.
35. S. Y. Wu, W. S. Huang, I. J. Chopra, M. Jordan, D. Alvarez and F. Santini, *Am. J. Physiol.*, 1995, **268**, E572–E579.
36. K. Moriyama, T. Tagami, T. Akamizu, T. Usui, M. Saijo, N. Kanamoto, Y. Hataya, A. Shimatsu, H. Kuzuya and K. Nakao, *J. Clin. Endocrinol. Metab.*, 2002, **87**, 5185–5190.
37. R. A. Heimeier, B. Das, D. R. Buchholz and Y. B. Shi, *Endocrinology*, 2009, **150**, 2964–2973.
38. M. Alaee, P. Arias, A. Sjodin and A. Bergman, *Environ. Int.*, 2003, **29**, 683–689.
39. D. Zalko, C. Prouillac, A. Riu, E. Perdu, L. Dolo, I. Jouanin, C. Canlet, L. Debrauwer and J. P. Cravedi, *Chemosphere*, 2006, **64**, 318–327.
40. U. M. Schauer, W. Volkel and W. Dekant, *Toxicol. Sci.*, 2006, **91**, 49–58.
41. S. Kitamura, N. Jinno, S. Ohta, H. Kuroki and N. Fujimoto, *Biochem. Biophys. Res. Commun.*, 2002, **293**, 554–559.
42. I. A. Meerts, J. J. van Zanden, E. A. Luijks, I. van Leeuwen-Bol, G. Marsh, E. Jakobsson, A. Bergman and A. Brouwer, *Toxicol. Sci.*, 2000, **56**, 95–104.
43. S. Iwamuro, M. Sakakibara, M. Terao, A. Ozawa, C. Kurobe, T. Shigeura, M. Kato and S. Kikuyama, *Gen. Comp. Endocrinol.*, 2003, **133**, 189–198.

44. O. Jagnytsch, R. Opitz, I. Lutz and W. Kloas, *Environ. Res.*, 2006, **101**, 340–348.
45. D. Melzer, N. Rice, M. H. Depledge, W. E. Henley and T. S. Galloway, *Environ. Health Perspect.*, 2010, **118**, 686–692.
46. J. G. Vos, E. Dybing, H. A. Greim, O. Ladefoged, C. Lambre, J. V. Tarazona, I. Brandt and A. D. Vethaak, *Crit. Rev. Toxicol.*, 2000, **30**, 71–133.
47. R. J. Witorsch, *Regul. Toxicol. Pharmacol.*, 2002, **36**, 118–130.
48. G. G. Kuiper, J. G. Lemmen, B. Carlsson, J. C. Corton, S. H. Safe, P. T. van der Saag, B. van der Burg and J. A. Gustafsson, *Endocrinology*, 1998, **139**, 4252–4263.
49. G. D. Charles, C. Gennings, B. Tornesi, H. L. Kan, T. R. Zacharewski, B. Bhaskar Gollapudi and E. W. Carney, *Toxicol. Appl. Pharmacol.*, 2007, **218**, 280–288.
50. M. Sole, D. Raldua, F. Piferrer, D. Barcelo and C. Porte, *Comp. Biochem. Physiol. C, Toxicol. Pharmacol.*, 2003, **136**, 145–156.
51. B. Quinn, F. Gagne, M. Costello, C. McKenzie, J. Wilson and C. Mothersill, *Aquat. Toxicol.*, 2004, **66**, 279–292.
52. R. Chaube, S. Mishra and R. K. Singh, *Toxicol. In vitro*, 2010, **24**, 1899–1904.
53. L. Sax, *Environ. Health Perspect.*, 2010, **118**, 445–448.
54. O. P. Heemken, H. Reincke, B. Stachel and N. Theobald, *Chemosphere*, 2001, **45**, 245–259.
55. Y. de Lafontaine, N. L. Gilbert, F. Dumouchel, C. Brochu, S. Moore, E. Pelletier, P. Dumont and A. Branchaud, *Sci. Total Environ.*, 2002, **298**, 25–44.
56. L. J. Guillette, Jr., *Environ. Health Perspect.*, 2006, **114**(suppl. 1), 9–12.
57. J. Legler, J. L. M. Broekhof, A. Brouwer, P. H. Murk, P. T. Saag, D. A. Vethaak, P. Wester, D. Zivkovic and B. Van Der Burg, *Environ. Sci. Technol.*, 2000, **34**, 4439–4444.
58. H. Chen, J. Hu, J. Yang, Y. Wang, H. Xu, Q. Jiang, Y. Gong, Y. Gu and H. Song, *Aquat. Toxicol.*, 2010, **96**, 53–61.
59. Z. Zeng, T. Shan, Y. Tong, S. H. Lam and Z. Gong, *Environ. Sci. Technol.*, 2005, **39**, 9001–9008.
60. K. Kurauchi, Y. Nakaguchi, M. Tsutsumi, H. Hori, R. Kurihara, S. Hashimoto, R. Ohnuma, Y. Yamamoto, S. Matsuoka, S. Kawai, T. Hirata and M. Kinoshita, *Environ. Sci. Technol.*, 2005, **39**, 2762–2768.
61. T. Ueno, S. Yasumasu, S. Hayashi and I. Iuchi, *Mech. Dev.*, 2004, **121**, 803–815.
62. S. R. Blechinger, J. T. Warren, Jr., J. Y. Kuwada and P. H. Krone, *Environ. Health Perspect.*, 2002, **110**, 1041–1046.

Application of Reporter Animals as Novel Tools in Food Safety Research

BALAJI RAMACHANDRAN AND ADRIANA MAGGI

Center of Excellence on Neurodegenerative Diseases and Department of Pharmacological Sciences, University of Milan, Via Balzaretti 9, I-20123 Milan, Italy

9.1 Introduction

It is becoming increasingly clear that various toxic agents which contaminate soil and water are also present in the vegetables and meats that we consume daily. The period of exposure to several of these compounds may be very long, raising the risk that they accumulate in our bodies to toxic levels. Such phenomena have been widely demonstrated, in particular in the case of endocrine disruptors (EDs), chemicals which mimic endogenous hormones (such as estrogens, androgens and thyroid hormones) by binding to their receptors and thereby altering their physiological signalling. Such a systemic imbalance may exert a significant impact on the entire organism and, indeed, several EDs were discovered because of their visible effects on the fertility of wild animals.[1,2]

At present, the methodologies available for the risk assessment of exposure to low doses of toxic compounds for prolonged periods of time are extremely limited. Although able to describe in detail the effects of drugs or chemicals on cells or tissues, *in vitro* procedures involving cell cultures cannot provide an estimate of human risk.[3-5] On the other hand, most animal models focus on the

Issues in Toxicology No. 11
Hormone-Disruptive Chemical Contaminants in Food
Edited by Ingemar Pongratz and Linda Vikström Bergander
© Royal Society of Chemistry 2012
Published by the Royal Society of Chemistry, www.rsc.org

toxicity of a compound towards a single organ (*e.g.* the uterotrophic assay for estrogenic compounds) and require a long time and the sacrifice of numerous experimental animals to demonstrate the extent of the undesirable effects.

Present procedures are also static in nature, revealing the consequences of exposure for a fixed period of time, without taking into account accumulation of the toxic xenobiotic. Moreover, the ability of most conventional tests to predict toxicity is weakened by the fact that they do not measure the direct activity of the toxicant on its molecular targets, but rather physiological end-points which may vary considerably between strains and species.[6–8] Thus, the findings of tests designed to measure acute and chronic toxicity cannot be easily extrapolated to humans.[9] The current strategy for overcoming these limitations is to use a panel of different tests, which, of course, dramatically increases the cost and time required to perform the analysis.

With regard to endocrine disruptors, several toxicology programmes have been designed and peer review panels set up to evaluate the validity of the tools currently available for studying the developmental and reproductive toxicity of low, no adverse effect levels (NOELs) present in food. For example, bisphenol A, diethylstilbestrol, estradiol, genistein, methoxychlor, nonylphenol and vinclozolin have been shown to exert significant effects after prolonged exposure to low concentrations.[10]

Novel technologies, including toxicogenomics and the other -omics provide new tools for risk assessment.[6] For example, RNA and protein expression profiles provide a preliminary indication of changes in gene expression in connection with toxicological evaluations.[11] The disadvantages of such approaches include the invasive surgery required to obtain the tissue sample, as well as the need for specialized equipment and expertise in gene expression profiling. Although such "-omics" approaches can help to reveal underlying molecular responses to chemicals or drugs and to predict the toxicity of unknown agents, in our opinion, more pragmatic strategies should be applied to mechanistic and predictive toxicology.[12] Progress in the development of *in vitro* and *in silico* methods during the last two or three decades has been enormous, but their application to risk assessment is still limited, due to their questionable relevance to humans.[13,14]

A major concern is that the information generated by current investigational tools in the field of toxicogenomics is not sufficiently detailed concerning the tissue/cellular localization or the time-point during exposure of the effects, most often representing only a single snapshot in time, since repeated invasive surgery is required to monitor the progression of a pathological process. In addition, the sample is destroyed during the analysis itself. New test systems that monitor the effects of bioactive molecules as they interact with endogenous receptors, ion channels, *etc.*, would help to reveal when the toxic events occur, as well as their subsequent progression.[15]

In this context, biomarkers can provide information concerning the dynamics of responses to xenobiotics and the mechanism of action, as well as offering insights of value for safety assessment (*e.g.* tissue distribution).[16] For example, reporter mice designed to evaluate the activities of intracellular receptors [*e.g.* ERs (estrogen receptors), ARs (androgen receptors) and THRs (thyroid hormone receptors)] could serve as innovative tools for the discovery and development of drugs, as well as for monitoring EDs present in the food chain or in the environment.[17,18] Such reporter mice allow the evaluation of receptor activation *in vivo* in virtually all organs/tissues during periods of acute or chronic exposure/treatment and provide systems that enable the study of target (*e.g.* receptor) activation in response to drug/contaminants to be monitored over periods of time, thereby providing more accurate information concerning pharmacodynamics and toxicity.

The aim of the present chapter is to demonstrate the applicability and utility of reporter animals in connection with risk assessment of EDs, showing how these models can be employed for the systematic *spatio*-temporal analysis of the activity of a given endogenous receptor and adapted for the study of selected xenobiotics administered acutely or for prolonged periods of time. The paradigmatic reporter system utilized was the ERE-*Luc* mouse generated in-house.

9.2 Materials and Methods

9.2.1 Experimental Animals and Treatments

Animal maintenance, handling and treatments were performed in accordance with European legislation and the experimental setting was approved by the Italian Ministry of Health.

9.2.1.1 *Anesthesia for In vivo Imaging*

Owing to the low levels of photon emission and the necessity to carry out quantitative imaging for at least 5 min, the mice had to be anesthetized, which was accomplished with a single s.c. injection of 50 μL of a solution of ketamine and xilazine in water (78:15:7). In this manner, mice can be immobilized for a period of 30 min, after which they recover very well if maintained at room temperature (23 °C). This dosage of anesthetic is not toxic if administered not more than three times in a 24-hour period or once every day for an extended period of time (*e.g.* 3 months).

9.2.1.2 *The Dietary Regimen Employed for the Evaluation of ER Activity*

Since even food containing low levels of phytoestrogens may activate ERs,[19] our animals were first maintained on a estrogen-free regular diet (4RF21, Mucedola) and then switched to AIN-93M diet (Mucedola) at least 96 hours prior to treatment. This AIN-93M diet was selected for low phytoestrogen content from a number of different diets on the basis of its relatively limited induction of luciferase activity in fasted ERE-Luc mice (not shown). Aliquots of the same batch of this diet were stored under vacuum at 4 °C until use.

Measurement of food and water consumption showed that 70% of the water and 75% of the food were consumed between 7:00 p.m. and 7:00 a.m., with peak consumption at around 1:00 a.m. Each animal was weighed on the first and last days of each experiment.

9.2.1.3 Administration of Cadmium Chloride

$CdCl_2$ was dissolved in physiological saline and administered as a single bolus (100 µL; 1 µg kg^{-1}) by oral gavage once every 24 h, immediately after each imaging session.

9.2.1.4 Administration of 17β-Estradiol, Genistein and Daidzein

Stock solutions of estradiol (E_2), genistein and daidzein were prepared by dissolving the compounds in 99% v/v ethanol to obtain a concentration of 10^{-2} M and were stored in glass vials at –20 °C in the dark. For gavage, these compounds were first diluted in physiological saline, the resulting solution vortexed thoroughly to ensure homogeneity and a single bolus of 100 µL then administered. In the case of chronic studies with estradiol, silastic pellets were used to release this hormone at physiological levels of 5 µg kg^{-1} d^{-1}.

Chronic treatments were carried out once every 24 h, immediately following each imaging session. The dose of genistein given was 5 mg kg^{-1} d^{-1}. In the case of acute experiments, genistein and daidzein were added to the mixture, mimicking soy milk, at a ratio of 3:1.[20] Owing to its reduced solubility, the volume of lyophilized soy milk administered was increased to 900 µL (given as three administrations of 300 µL each at 45 min intervals).

9.2.2 Administration and Distribution of the Luciferin Substrate

In all cases, saturating doses of the substrate luciferin were employed, as determined earlier.[21]

9.2.3 *In Vivo* and *Ex Vivo* Bioluminescence Imaging

For *in vivo* measurement of bioluminescence, the mice were anesthetized by subcutaneous injection of a solution of ketamine (78 mg kg^{-1}) and xilazine (6 mg kg^{-1}), after which the luciferase substrate luciferin (25 mg kg^{-1}) was administered intraperitoneally and 20 min later the bioluminescence was quantified. *Ex vivo* assessment of bioluminescence was performed on the final day of the experiment, immediately after the last *in vivo* imaging. For this purpose, selected organs (the liver, small intestine, kidneys and brain) were excised and placed in a petriplate coated with phosphate-buffered saline (PBS).

In both cases, bioluminescence was quantified employing a Night Owl imaging unit (Berthold Technologies, Bad Wildbad, Germany) consisting of a cooled, charge-coupled slow-scan Peltier camera equipped with a 25 mm/f 0.95

lens and operated by WinLight software (Berthold Technologies). For measurement of photon emission, the anesthetized mice or dissected organs were placed in a light-tight chamber, where gray-scale images were first obtained using dim light, and photon emission then registered for 5 or 15 min, respectively. Merging of the pictures enabled localization of the emission signal to specific organs or tissues (in the figures shown, the luciferase signal is depicted as pseudo-colors: blue when low and white when high). For quantification, photon emission in the regions of interest was counted and the signals thus obtained were integrated from each anatomical area. Photon emission was defined as the number of counts per second per centimetre square (Cts s^{-1} cm^{-2}) and monitored with WinLight32 imaging software (Berthold Technologies). Normalization was performed by utilizing an external source of photons to assess the instrumental efficiency of counting (Glowell, Luxbiotech, Edinburgh, UK).

9.2.4 Bread-based Rodent Diets

Commercially available white wheat toast (expected to contain low levels of Cd) or white wheat toast supplemented with flaxseed (expected high level of Cd) was used for the study. Highest feasible amount of bread (42.92 %) was added to the semi-synthetic AIN93G rodent diet without compromising the macronutrient composition or pellet quality of the diets. The Cd levels in these breads were analysed in duplicate with graphite furnace atomic absorption spectrometry (GF-AAS) using the standard addition quantification method as described previously.[22]

9.3 Results and Discussion

9.3.1 Generation and Characterization of the ERE-*Luc* Reporter Mice

The ERE-*Luc* strain of mice was designed for rapid, quantitative and systemic analysis of the activity of ERs *in vivo* with and without exogenous estrogenic stimuli.[23,24] In brief, this mouse strain was generated by integrating into its genome a transgene containing a wild-type firefly luciferase gene driven by a simple promoter containing a multimerized estrogen responsive element (ERE) to minimize position effects; the transgene was flanked by insulator sequences.[25] Several factors were taken into consideration: (i) the estrogen responsive gene had to be expressed systemically (for a comprehensive spatial analysis of ER activity);[23] (ii) the reporter gene had to encode short-life proteins (to reflect the dynamics of the ER action);[26] (iii) the expression of the reporter gene had to be directly proportional to the extent of transcriptional activation of the ER;[27] (iv) the sensitivity of the reporter system had to be sufficient to show physiological changes in ER activity;[27] (v) the activity of the reporter gene had to be in the same cells where the ER was expressed and had to be shown to occur where endogenous target genes were activated; and, of course, (vi) the reporter gene had to encode a nonmammalian protein, easily measured by quantitative assays and a good antigen to facilitate its cellular localization by immunohistochemistry and be applied to non-invasive imaging technologies.[28] We have demonstrated that all of these features are exhibited by the ERE-*Luc* model in our previous publications.

For this model to be applied to monitor the dynamics of ER activity upon interaction with the toxic substances reproducibly and quantitatively, several other conditions need to be fulfilled. First, it is important to verify that the substrate luciferin diffuses through all the cells of interest and attains a concentration sufficient to saturate the reporter luciferase; this can be achieved experimentally with a simple time-course and dose-response.[21] Next, the reporter must reflect the dynamics of ER transcriptional activity with a sensitivity sufficient to detect the changes induced by endogenous, physiological stimuli. As depicted in Figure 9.1A, the ERE-*Luc* model enables evaluation of the changes in ER activity occurring during the estrous cycle of the female mouse, thus following the physiological, oscillating synthesis of estrogens, by measuring photon emission in selected regions of interest (RoI), such as liver and vagina (Figure 9.1B). Ovariectomy significantly reduces ER signalling in the female mice and very little ER activity is observed in male animals (Figure 9.1C).

9.3.2 Application of ERE-*Luc* Mice to Monitoring Tissue-specific Effects of Estrogenic Compounds over Space and Time

One of the advantages of *in vivo* imaging methodologies is that the study can be carried out in a single animal where the activity of the reporter can be monitored longitudinally upon exposure to the environmental factors or diet of interest. Although the major limitation of the use of bioluminescence is associated with the impossibility of a reliable measurement in the three dimensions, a well standardized analysis of the photon emission in selected RoI may still provide a wealth of information, provided that the measurement is done in a very consistent and reproducible manner. To obtain the necessary reproducibility, animals were placed in a cast ensuring the proper positioning of the animal and, to generate a well-ordered/reproducible collection of data sets, we designed a segmentation algorithm to automatically analyze photon emission from the body areas of interest.[29]

With this system we monitored the ER activity in ovariectomized (OVX) female ERE-*Luc* mice administered chronically with 17β-estradiol (the pellet implanted in the neck of the animals released 5 μg kg^{-1} d^{-1}). As shown in Figure 9.2A, the analysis of long-term exposure to estrogens showed a differential response to the treatment of the two organs. In the vagina the ER became fully activated a few days after the initiation of the treatment and its activity was maintained high for the entire length of the study. However, in the liver, ER activity increased dramatically during the first day of treatment, following which the bioluminescence decreased progressively with time. This observation underlines the existence of a diverse activity of the same ER ligand in different tissue, as predicted by molecular studies demonstrating that the influence of ER on transcriptional activity is regulated in a tissue-specific manner by

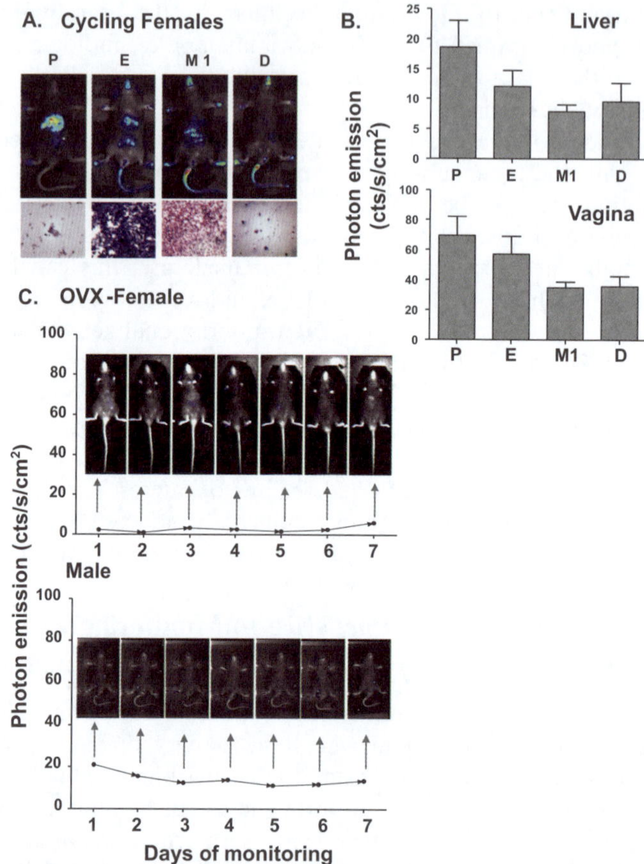

Figure 9.1 Monitoring ER activity during the estrous cycle in ERE-*Luc* mice. (A) Alterations in ER activity during the estrous cycle in female mice can be easily monitored. P, proestrous; E, estrous; M1, metestrous 1; D, diestrous. (B) Photon emission measured on liver and vagina of female mice during different phases of the estrous cycle. (C) Bioluminescence in ovariectomized (OVX) female and male ERE-*Luc* mice was measured daily for a week by bioluminescence imaging.

co-regulators, as well as by the observed effects of ligands such as the selective estrogen receptor modulators (SERMs).[30]

Most interestingly, the analysis of bioluminescence in the OVX control animals showed that even in the absence of exogenous stimuli the ER in OVX mice appears to undergo waves of activity, presumably due to factors other than estrogens produced by the ovaries or other organs.

Focusing on this phenomenon, we discovered that ER-dependent transcriptional activity can be regulated tissue-specifically in the absence of the endogenous ligand *via* several different mechanisms. For instance, consumption of non-estrogenic food enhances hepatic ER activity selectively.[19] Moreover,

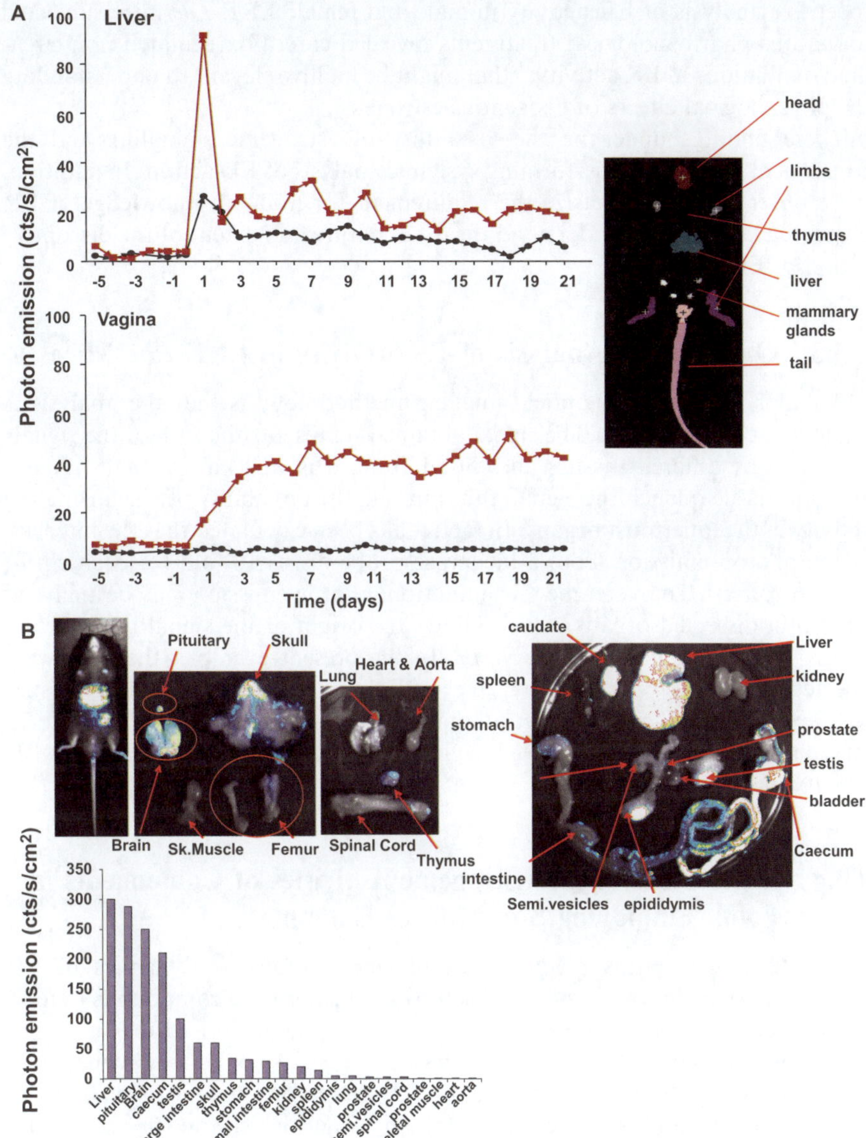

Figure 9.2 Bioluminescence imaging of ERE-*Luc* mice treated with estradiol for a period of 21 days. (A) Quantitative analysis of photon emission from the liver and vagina of estradiol-treated and control mice. Data represent the average of determinations made in groups of six animals each. (B) *Ex vivo* imaging of the organ-specific expression of luciferase. Photon emission was initially recorded in the whole animal and then the suspected tissues were dissected and *ex vivo* imaging analysis was measured in isolated tissues to identify the tissue-specific localization of the luciferase signal. The emitted photons are counted and expressed as Cts s^{-1} cm^{-2}.

repetitive analysis of ER activity in male and female ERE-*Luc* mice subjected to various pharmacological treatments revealed carefully regulated non-circadian oscillations in ER activity[31] that might be highly relevant to understanding the physiological effects of exogenous estrogens.

These findings underline the versatility of estrogenic signalling and the consequent necessity of performing systemic analysis of ED action. In addition, these observations emphasize the requirement for in-depth knowledge of ER physiology in order to understand the exact nature and extent of the disruptive action of EDs.

9.3.3   Quantitative Analysis of ER Activity in ERE-*Luc* Mice

The major limitation of optical imaging methodology is that the analysis is made two-dimensionally. The signal obtained represents the sum of the signals generated by different tissues and, in addition, it is well known that different layers of tissue quench the signal, thus limiting the possibility of visualizing the activity in the innermost organs. In fact, it has been calculated that the intensity of the photon emission from a luciferase probe drops 10-fold for each cm of tissue traversed. To overcome this limitation, photon emission may be analyzed *ex vivo* in dissected organs, which allows the origin of the signal in the living animal to be identified (Figure 9.2B). In the present example, the mice were injected i.p. with 50 µg kg^{-1} of 17β-estradiol; 6 hours later, the target tissues were excised (the luciferin substrate having been injected i.p. 15 min prior to imaging) and subjected to *ex vivo* imaging to obtain a reliable assessment of ER activity in inner organs.[32]

9.3.4   Assessment of the Estrogenic Activities of Components in Food, Employing Soy Milk as Example

Having established this methodology for monitoring the dynamics of ER action, we tested its validity for the detection of estrogenic compounds in food in two sets of experiments. Firstly, the estrogenic potential of a food substance known to contain high levels of natural estrogens, such as isoflavones, was examined; secondly, the ability of our ERE-*Luc* mouse model to evaluate an added food contaminant such as cadmium, which is reported to exert mild estrogenic action, was determined. The protocol employed, took into account all possible sorts of interference that might hinder the detection of small responses to low doses of toxicants. Furthermore, this protocol was strictly standardized for conducting longitudinal studies.[33]

The acute induction of ER transcriptional activity upon oral administration of genistein (at a nutritional dose), soy milk and known soy isoflavones was examined in 4-month-old male ERE-*Luc* mice. The isoflavones in soy milk that are most well-characterized are genistein and daidzein,[20] although other compounds exhibiting estrogen-like activity but not yet fully characterized, such as glyceitin, may be present at lower concentrations.

Accordingly, a mixture of genistein and daidzein was prepared containing the amounts of these two compounds ingested (genistein 38 µg and daidzein 12.7 µg) as reported[20] to be present in a ratio of 3:1 that is normalized to the amount of soy milk consumed per day and 0.8 g of lyophilized soy milk (amount from one day milk consumption). As a positive control, E_2 (50 µg kg^{-1}) was administered at a pharmacological dosage known to fully activate the ERs. ERE-*Luc* male mice were treated by oral gavage with the above mixtures and the state of transcriptional activation of the ERs was measured by bioluminescence imaging (BLI) at 3, 6 and 17 h after treatment. Photon emission from selected body areas was measured as described in Section 9.2.3.

E_2 induced significant activation of ERs in all the body areas taken into consideration (Figure 9.3), with a peak of activity at 6 h (chest, abdomen and limb). This was in line with the previous observations by several reports,[27,34] where the hormone was administered either intraperitoneally or subcutaneously. After the treatment with genistein, the only area that was observed to change in photon emission was the chest, with an increased photon emission that was gradual in time and was highest 6 h after gavage.

Thus, genistein administered orally at a dose of 5 mg kg^{-1} activated ER activity in mouse liver to approximately 10% of the response obtained with 50 µg kg^{-1} estradiol. This was expected, since it is well established[4] that the potency of genistein is lower than E_2[35] in this respect and because of the relatively high dose of E_2 used in the experiment. Further, in this time-course study, a significant increase of photon emission was observed in chest and abdomen 6 h after the treatment with lyophilized soy milk, but the soy milk-mimic solution did not have significant effects on ER activity.

These experiments demonstrate, for the first time, the efficacy of soy milk in activating ERs. Moreover, comparison of the corresponding responses to purified genistein and daidzein or in combination revealed that the effect of soy milk was not simply due to the summation of the effects of the estrogenic compounds known to be present in soy, but some other factors present in the food matrix synergized with the known estrogenic compounds must account for the potent activation of ER activity both in short-term as well as in long-term studies.[33]

9.3.5 Administration of Bread Contaminated with Cadmium Causes Weak Estrogenic Effects in Adult Female ERE-*Luc* Mice

Cadmium is a well-known toxic compound present in the general environment due to industrial waste.[36] Human exposure to cadmium occurs mainly because of its presence in seeds and plants. For example, wheat is a staple food; daily ingestion of wheat bread, particularly bread enriched with specific seeds, may be the source of continuous accumulation of this metal and is slowly excreted from the body. At present, it is very difficult to determine the exact dosage at which this metal becomes toxic or even to detect its presence in the food chain.

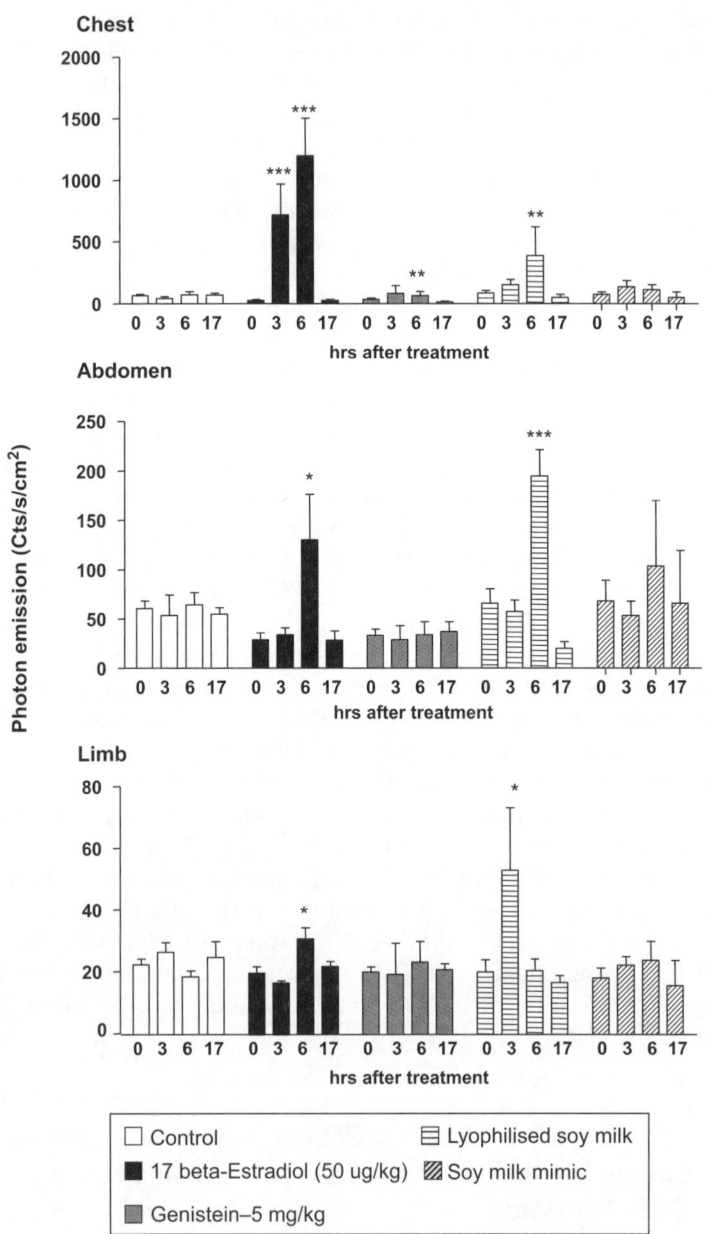

Figure 9.3 Acute effects of soy milk and its isoflavones on ER activity. Photon emission was initially recorded in the whole animal at 0, 3, 6 and 17 h after treatment. Data represent mean $\pm$ SEM ($n = 5$) and asterisks indicate a significant difference from control, as assessed by two-way ANOVA plus Bonferroni *post hoc* test: *, $p < 0.05$, **, $p < 0.01$, ***, $p < 0.001$.

Because cadmium has been reported to exert estrogenic activity, we tested in female ERE-*Luc* mice the extent to which a prolonged exposure to a bread-based food prepared with contaminated wheat could activate the ER in different organs. Appropriate diets were prepared as described in the methodology section and mice were fed with this diet for 21 days. In parallel, we treated by gavage additional groups of mice with $CdCl_2$ and E_2 to evaluate the contribution of inorganic cadmium alone to the effects observed and the extent of perturbation of the ER signalling system compared with the natural hormone.

Figure 9.4 depicts photon emission due to luciferase activity on different days during the 21 days experimental time and indicates that E_2 caused the

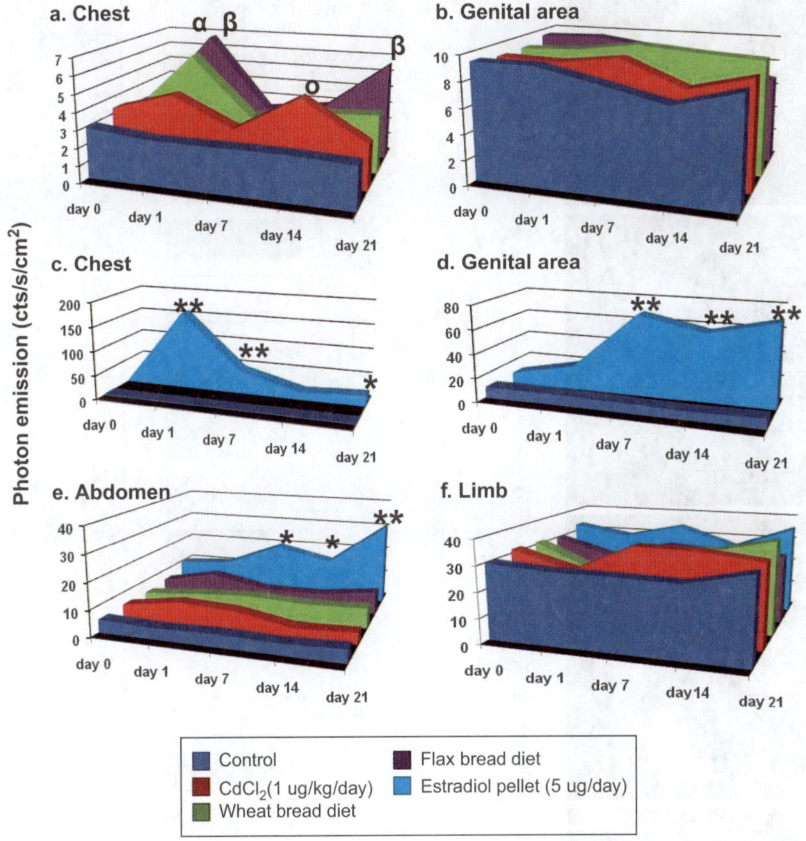

Figure 9.4 The effects of bread diets and cadmium on ER activity in various areas of the body of OVX female ERE-*Luc* mice. The animals were subjected to bioluminescence imaging during baseline, 1, 7, 14 and 21 days of treatment and the data expressed as Cts s^{-1} cm^{-2}. Data represents mean $\pm$ SEM ($n = 5$) and asterisks indicate a significant difference from control, as assessed by a two-way ANOVA plus Bonferroni *post hoc* test: *, $p < 0.05$; **, $p < 0.01$ (E_2-treated mice *versus* controls); °, $p < 0.05$ ($CdCl_2$-treated mice *versus* controls); α, $p < 0.05$ (wheat bread diet treated mice *versus* controls); β, $p < 0.05$ (flax bread diet treated mice *versus* controls).

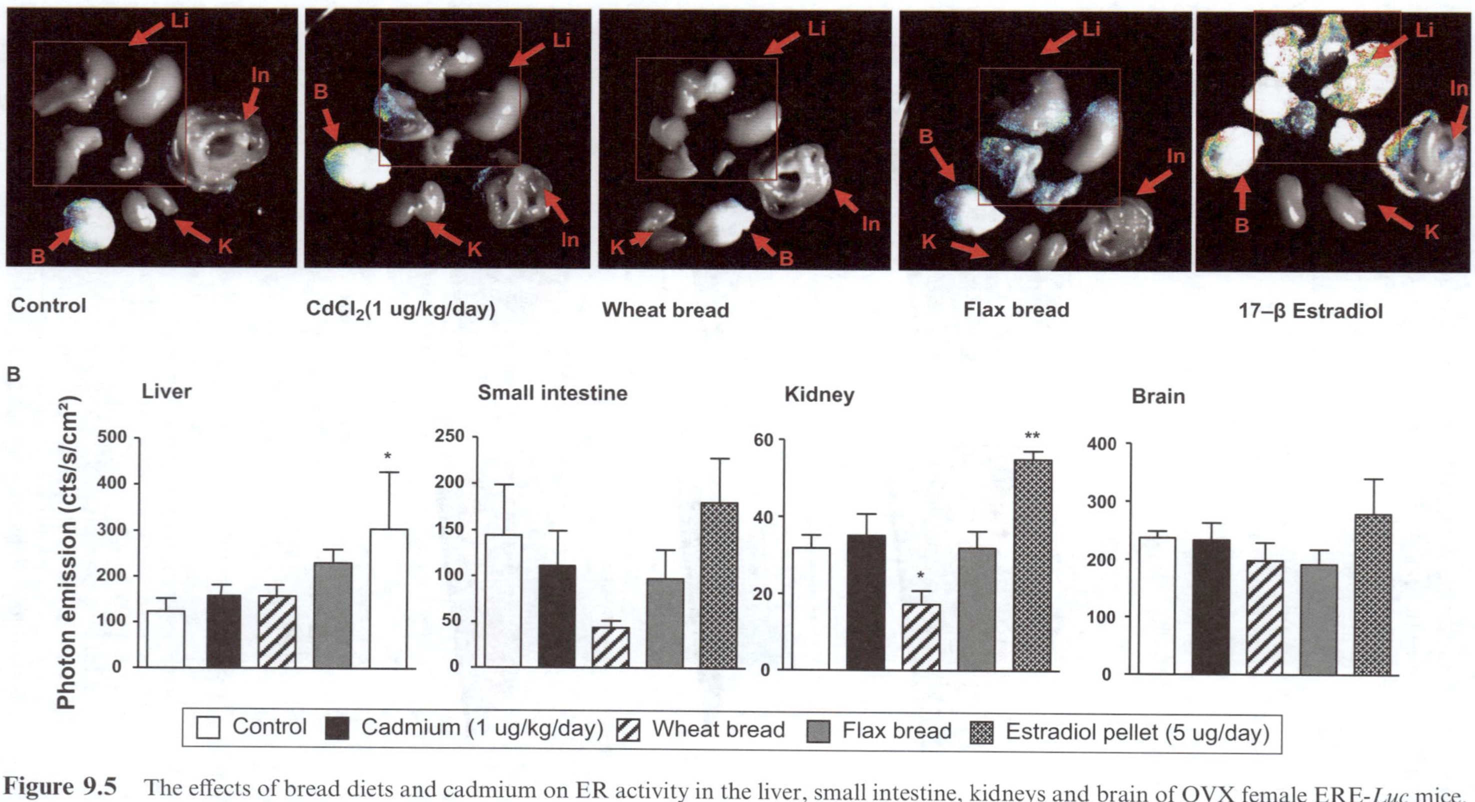

Figure 9.5 The effects of bread diets and cadmium on ER activity in the liver, small intestine, kidneys and brain of OVX female ERE-*Luc* mice. *Ex vivo* imaging was performed on the final day of a 21-day treatment. (A) Representative images. Li, liver; In, small intestine; B, brain; K, kidney. (B) Quantification of the photon emission from the individual excised tissues. Data represent mean ± SEM ($n = 5$) and asterisks indicate a significant difference from control, as assessed by two-way ANOVA plus Bonferroni *post hoc* test: *, $p < 0.05$.

well-known changes in ER activity in the liver and reproductive area. The flax-enriched bread diet affected ER signalling in the liver significantly; most treatments appeared to have a slight effect on bone, but with the small number of experimental animals analyzed did not allow us to definitely prove that these effects were significant. In comparison to the controls, the luciferase activity in all treated groups had more pronounced fluctuations, indicating the possible presence of an unknown activator, which should be examined further. On the final day of treatment, the mice were euthanized and luciferase activity was measured in each specific organ by *ex vivo* imaging.

Although *ex vivo* analysis did not show a major alteration in ER activity, a mild inhibition of ER activity was observed in the wheat bread treated group, most significantly in the kidneys. The enhanced ER activity observed in the chest area of the group consuming flax bread was not shown in the liver, indicating that the signal from the chest area did not originate from this organ (Figure 9.5).

9.4 Conclusions

The reporter mouse technology combined with bioluminescence imaging represents a novel and highly sensitive tool for toxicological studies. The advantage of this approach over classical toxicology is that it allows non-invasive and quantitative measurement of the molecular events in response to a toxicant. An important feature of this technological platform is the possibility of performing repeated observations on the same animal, which minimizes the variability and allows characterization of the effects of even low doses of a chemical compound that bioaccumulates in time, leading to potential physiological perturbations. Thus subtle perturbations at the physiological homeostasis and bioaccumulative toxicants can be assessed, which are outstanding advantages over -omics and cell-based studies.

The present investigation illustrates the clear advantage of the utilization of reporter mice for the rapid identification of the tissues and organs which are affected by dietary endocrine disruptors, even at low doses. The introduction of a surrogate marker (ERE-driven luciferase expression) provides a global view of the tissues influenced by the compound of interest, unequivocal determination of the dosage and the period of exposure required to alter the activity of the receptor, and longitudinal monitoring of a single animal during repeated exposure.

It should be emphasized that the ERE-*Luc* model provides for the first time a clear picture of the intricacies of ER signalling. Indeed, such a complex pattern is expected because of the major role played by ERs in reproduction, a function which from the phylogenetic point of view is extremely well controlled by environmental and dietary factors. We are convinced that the limited knowledge concerning ER physiology that is presently available hampers our understanding of the exact, long-term consequences of tissue-specific perturbation of ER signalling and, therefore, our interpretation of fluctuations in this activity with time. Systematic analysis of tissue-specific ER behaviour under pathophysiological conditions and in response to xenobiotics[31] will soon allow us to draw more definitive conclusions regarding the ER behaviour required to maintain a healthy status.

The emerging complexity of ER activity, which is certainly shared by other target receptors for ED, underscores the major limitations of current approaches to the study of ED toxicity and provides a glimpse of methodologies that are potentially more fruitful. Moreover, outside the complex field of ED, molecular imaging should be widely applicable to the evaluation of the *in vivo* effects of numerous endogenous, dietary and environmental toxicants, as demonstrated in a series of previous reports.[19,37–40]

References

1. L. J. Guillette, *Cent. Eur. J. Public Health*, 2000, **8** (suppl), 34–35.
2. M. R. Milnes, D. S. Bermudez, T. A. Bryan, T. M. Edwards, M. P. Gunderson, I. L. V. Larkin, B. C. Moore and L. J. Guillette, *Environ. Res.*, 2006, **100**, 3–17.
3. J. P. Nash, D. E. Kime, L. T. M. V. der Ven, P. W. Wester, F. Brion, G. Maack, P. Stahlschmidt-Allner and C. R. Tyler, *Environ. Health Perspect.*, 2004, **112**, 1725–1733.
4. G. D. Charles, *ILAR J.*, 2004, **45**, 494–501.
5. R. L. Brent, *Pediatrics.*, 2004, **113** (suppl. 4), 984–995.
6. L. L. Smith, *Trends Pharmacol. Sci.*, 2001, **22**, 281–285.
7. S. Kacew and M. F. Festing, *J. Toxicol. Environ. Health*, 1996, **47**, 1–30.
8. P. Li, *Chem. Biol. Interact.*, 2004, **150**, 3–7.
9. A. Secchi and V. Deligianni, *Curr. Opin. Allergy Clin. Immunol.*, 2006, **6**, 367–372.
10. NIEHS, National Toxicology Program (NTP), *Endocrine Disruptors Low Dose Peer Review Report*, National Institute of Environmental Life Sciences, Research Triangle Park, NC, 2001, pp. 1–467.
11. J. M. Naciff and G. P. Daston, *Toxicol. Pathol.*, 2004, **32** (suppl 2), 59–70.
12. M. R. Fielden and T. R. Zacharewski, *Toxicol. Sci.*, 2001, **60**, 6–10.
13. B. J. Blaauboer, *Toxicol. Lett.*, 2008, **180**, 81–84.
14. T. Seidle and M. L. Stephens, *Toxicol. In Vitro.*, 2009, **23**, 1576–1579.
15. B. Hock and M. Seifert, *Nat. Biotechnol.*, 1999, **17**, 1162–1163.
16. P. S. Murphy, T. J. McCarthy and A. S. K. Dzik-Jurasz, *Br. J. Radiol.*, 2008, **81**, 685–692.
17. J. Muncke, *Sci. Total Environ.*, 2009, **407**, 4549–4559.
18. E. L. Teuten, J. M. Saquing, D. R. U. Knappe, M. A. Barlaz, S. Jonsson, A. Björn, S. J. Rowland, R. C. Thompson, T. S. Galloway, R. Yamashita, D. Ochi, Y. Watanuki, C. Moore, P. H. Viet, T. S. Tana, M. Prudente, R. Boonyatumanond, M. P. Zakaria, K. Akkhavong, Y. Ogata, H. Hirai, S. Iwasa, K. Mizukawa, Y. Hagino, A. Imamura, M. Saha and H. Takada, *Philos. Trans. R. Soc. London, Ser. B*, 2009, **364**, 2027–2045.
19. P. Ciana, A. Brena, P. Sparaciari, E. Bonetti, D. D. Lorenzo and A. Maggi, *Endocrinology*, 2005, **146**, 5144–5150.

20. E. A. P. Nahas and J. Nahas-Neto, *Curr. Drug Therapy*, 2006, **1**, 31–36.
21. Biserni, F. Giannessi, A. F. Sciarroni, F. M. Milazzo, A. Maggi and P. Ciana, *Mol. Pharmacol.*, 2008, **73**,1434–1443.
22. G. Khan, P. Penttinen, A. Cabanes, A. Foxworth, A. Chezek, K. Mastropole, B. Yu, A. Smeds, T. Halttunen, C. Good, S. Mäkelä and L. Hilakivi-Clarke, *Reprod. Toxicol.*, 2007, **23**, 397–406.
23. P. Ciana, G. D. Luccio, S. Belcredito, G. Pollio, E. Vegeto, L. Tatangelo, C. Tiveron and A. Maggi, *Mol. Endocrinol.*, 2001, **15**, 1104–1113.
24. A. Maggi, L. Ottobrini, A. Biserni, G. Lucignani and P. Ciana, *Trends Pharmacol. Sci.*, 2004, **25**, 337–342.
25. G. Ashcroft *et al.* (First European Therapeuticals), WO Pat. Appl., PCT/EP2006007488.
26. J. F. Thompson, L. S. Hayes and D. B. Lloyd, *Gene*, 1991, **103**, 171–177.
27. P. Ciana, M. Raviscioni, P. Mussi, E. Vegeto, I. Que, M. G. Parker, C. Lowik and A. Maggi, *Nat. Med.*, 2003, **9**, 82–86.
28. C. H. Contag, S. D. Spilman, P. R. Contag, M. Oshiro, B. Eames, P. Dennery, D. K. Stevenson and D. A. Benaron, *Photochem. Photobiol.*, 1997, **66**, 523–531.
29. G. Rando, E. Casiraghi, S. Arca, P. Campadelli and A. Maggi, in *Proceedings of the 10th European Congress of the International Society for Stereology*, Bologna, Italy, 2009, p. 60.
30. E. K. Shanle and W. Xu, *Adv. Drug Deliv. Rev.*, 2010.
31. G. Rando, D. Horner, A. Biserni, B. Ramachandran, D. Caruso, P. Ciana, B. Komm and A. Maggi, *Mol. Endocrinol.*, 2010, **24**, 735–744.
32. A. Stell, S. Belcredito, P. Ciana and A. Maggi, *Mol. Imaging*, 2008, **7**, 283–292.
33. G. Rando, B. Ramachandran, M. Rebecchi, P. Ciana and A. Maggi, *Toxicol. Appl. Pharmacol.*, 2009, **237**, 288–297.
34. J. G. Lemmen, R. J. Arends, A. L. van Boxtel, P. T. van der Saag and B. van der Burg, *J. Mol. Endocrinol.*, 2004, **32**, 689–701.
35. G. G. Kuiper, B. Carlsson, K. Grandien, E. Enmark, J. Häggblad, S. Nilsson and J. . Gustafsson, *Endocrinology*, 1997, **138**, 863–870.
36. M. Faria, D. Huertas, D. X. Soto, J. O. Grimalt, J. Catalan, M. C. Riva and C. Barata, *Chemosphere*, 2010, **78**, 232–240.
37. D. D. Lorenzo, R. Villa, G. Biasiotto, S. Belloli, G. Ruggeri, A. Albertini, P. Apostoli, M. Raviscioni, P. Ciana and A. Maggi, *Endocrinology*, 2002, **143**, 4544–4551.
38. P. Mussi, P. Ciana, M. Raviscioni, R. Villa, S. Regondi, E. Agradi, A. Maggi and D. D. Lorenzo, *Brain Res. Bull.*, 2005, **65**, 241–247.
39. M. Penza, C. Montani, A. Romani, P. Vignolini, P. Ciana, A. Maggi, B. Pampaloni, L. Caimi and D. D. Lorenzo, *Toxicol. Sci.*, 2007, **97**, 299–307.
40. D. D. Lorenzo, G. Rando, P. Ciana and A. Maggi, *Toxicol. Sci.*, 2008, **106**, 304–311.

In Silico *Approaches to Screening Dietary Endocrine Disruptors*

RODOLFO GONELLA DIAZA, ALESSANDRA RONCAGLIONI AND EMILIO BENFENATI

Institute for Pharmaceutical Researches "Mario Negri", Via La Masa 19, 20156 Milano, Italy

10.1 Introduction: Filling Gaps in Our Knowledge and Reducing the Use of Animals at the Same Time

Toxicity testing of chemicals could involve the use of many experimental animals and be quite costly. Accordingly, the general legislative trend has been to promote, wherever possible, the application of alternative methods for predicting the physicochemical and (eco)toxicological properties of xenobiotics for regulatory purposes.

An important and relatively novel alternative to *in vivo* testing is computer-based procedures, more commonly referred as *in silico* approaches. This strategy has produced predictive models that are more rapid and less costly than both *in vivo* and *in vitro* tests, allowing a large amount of data concerning numerous chemical substances to be generated in a short time without the use of experimental animals. *In silico* procedures are already accepted by regulatory agencies in the United States, Canada, Japan and certain EU countries (at national level) *in lieu* of experimental testing for a series of end-points. At the

Issues in Toxicology No. 11
Hormone-Disruptive Chemical Contaminants in Food
Edited by Ingemar Pongratz and Linda Vikström Bergander
© Royal Society of Chemistry 2012
Published by the Royal Society of Chemistry, www.rsc.org

pan-European level, initial steps towards the adoption and acceptance of *in silico* approaches have been taken in the past few years. One important example of this is the regulation of cosmetics within the EU (Council Directive 76/768/EEC), which puts an end to animal testing by imposing a ban on testing finished cosmetic products (since September 2004) and ingredients (since March 2009) on animals. This regulation also imposes a ban on marketing finished cosmetic products which have been tested on animals, or which contain ingredients that have been tested on animals (since March 2009 with the exception of tests for repeated-dose toxicity, reproductive toxicity and toxicokinetics, which ban will apply from March 2013).

This ban forces the use of alternative procedures, including *in silico* methods.

EC regulation 1907/2006, entitled the *Registration, Evaluation, Authorisation of Chemicals* (REACH), is less restrictive, stating that all chemical substances produced or imported into the EU, in an amount greater than one tonne, must be registered. This registration entails compilation of a detailed dossier for each substance, including physicochemical and (eco)toxicological information. In order to reduce animal testing, the REACH regulation clearly foresees and promotes the use of alternative approaches, including *in silico* procedures.

Application of *in silico* methods to hormone-disrupting chemicals is particularly complicated for a number of reasons, including certain aspects of these methods themselves and the lack of a clear understanding of the biological and toxicological processes underlying endocrine-disrupting phenomena, as well as differences in the type of information requested (a category of activities, possible mechanisms, ranking of potency, *etc.*) and the purpose of this information (*e.g.* for regulatory purposes, research & development, *etc.*).[1] Here, we categorize and discuss these methods from a technical point of view, but it is important to keep in mind that other characteristics also influence the choice of a method. The user should have a clear understanding of both the information that can be obtained using these techniques, which we will clarify here, and the purposes for which they can be used.

10.2   *In Silico* Approaches

In the light of present legislative trends, both accepting and promoting the use of *in silico* approaches, it seems reasonable to apply these methods to the assessment of the endocrine-disrupting potency of chemicals in food. Such application of computational models assumes that the potency of an endocrine disruptor (ED) can be deduced from its structure. Such structure–activity relationships could be analyzed in two main ways: (1) by considering the ligand alone, disregarding the structure of the receptor; or (2) by simulating the interaction between the receptor binding site and the ligand.

The three main techniques presented here are: the quantitative structure–activity relationship (QSAR) and 3D-QSAR (both of which are employed for the first type of analysis) and virtual docking (which belongs to the second class).

10.2.1   QSAR Models: the General Approach

QSAR models are based on the idea that the properties of a given chemical compound are determined by its composition and structure.[2] It is quite obvious that if some part of the molecule is modified, its properties will change. The real challenge is to predict properties from a chemical structure and this could be very ambitious in cases where the properties are complex (*e.g.* when the role of the chemical structure is mediated by several steps and components). Certain simple physicochemical properties, such as the boiling point, can be relatively easily predicted from the chemical structure, whereas others, in particular complex biological properties such as endocrine-disrupting behaviour, are much more difficult to predict.

The three conceptual components of the QSAR model are: the chemical information provided by the structure, the property to be modelled and the mathematical function that links these two. A somewhat simplified version of this model is the so-called structure–activity relationship (SAR), in which, for example, the expected activity is related to the presence of a certain chemical moiety in the molecule. For example, if the molecule contains a nitrosamine group, it is expected to be carcinogenic. Accordingly, the output of the SAR model is categorical, *e.g.* toxic or not.

In contrast, the output of the QSAR model is a continuous parameter, such as a toxic dose. Recently, this approach has also been used to assess categorical properties, such as probability of belonging to a certain class of chemicals (*e.g.* carcinogenic), but still employing a mathematical model. For the sake of simplicity, we will discuss QSAR procedures in a general manner, including SAR approaches.

Other strategies, in some ways related to SAR and QSAR, have been utilized to relate the chemical structure of a molecule to its properties. For instance, the Threshold of Toxicological Concern (TTC) concept, proposed for application to food contaminants, also evaluates the presence of certain chemical moieties in the molecule. However, in this case the chemical moieties are clustered into a few (three or four) major categories, without taking into consideration the contribution of several different fragments to a specific toxic mechanism, as is typically done with SAR.[3] With this approach, each clustered group of chemicals is associated with a certain concentration, which represents a level of exposure supposed to be safe and hence considered as an acceptable level of contamination.

In another approach, referred to as read-across, the properties of the target compound are evaluated only by comparison with other compounds with known experimental values for the property under evaluation and sharing similarities with the target chemical. Thus, read-across is somehow a mini-QSAR (or mini-SAR) model. By grouping compounds into families, the hypothesis is that members of the same family share the same properties or possess an identifiable trend, although with certain differences due to differences in chemical structure. A key aspect of this approach is to select the most proper compounds since the concept of similarity is not so trivial.[4]

QSAR, read-across, grouping methods and more in general *in silico* non-testing methods are often referred, as well as *in vitro* testing, as alternative approaches to *in vivo* tests. This type of definition could be, in some cases, misleading. For instance, REACH legislation clearly states that all available information should be used for the compilation of chemicals' dossiers. However, REACH defines *in vitro* and *in silico* approaches, if available, as preferable ways to obtain information on chemical substances.

10.2.2 QSAR Models: the Chemical Information

The chemical information contained in the structure of a molecule can be described in many different ways. In the case of QSAR models, the many different possibilities can be simplified by stating that certain approaches focus on fragments present in the molecule, while others take into account the global properties of the compound.[5]

Fragments can be predefined on the basis of chemical residues and even a simple count of atoms. Another possibility is to list toxic fragments expected to exert a certain effect. A third approach is to derive the fragments of interest using the model itself, with no predefined list of fragments or even their nature. In this latter case the software typically generates fragments and then assesses whether any of these occurs more frequently in the toxic compounds.

In addition to the fragments responsible for the activity of interest, certain programs take into account other fragments, present in the same structure, which might modulate or inhibit this activity. Certain programs involve thousands of predefined fragments, thereby allowing rapid evaluation of new chemical structures. In many cases the pharmaceutical industry utilizes such lists of chemical fragments (also called structural keys) to process huge libraries of candidate drugs.

The other major approach involves chemical descriptors, of which there are now thousands and their number is continuously increasing, in part through the combination of several, simpler ones into a more complex descriptor. These chemical descriptors can be quite simple (*e.g.* the molecular weight) or much more complicated, requiring complex calculations. Certain descriptors are designed to describe interrelationships between several atoms and the overall chemical nature of the compound, while others involve the energy or polarizability of the substance. The classes of molecular descriptors employed most commonly in QSAR models are shown in Table 10.1.

A distinction is commonly made between bi- and tridimensional chemical descriptors. In the first case, information is derived from the simple bidimensional representation of the chemical compound; while in the other, a description of the Cartesian space (taking into account the conformation of the chemical, which can vary) is required.

10.2.3 QSAR Models: the Algorithm

The hypothesis underlying QSAR models, that the properties of a compound are determined by its chemical structure, can be expressed as an algorithm.

Table 10.1 The major classes of molecular descriptors employed in predictive QSAR models.

Descriptor class	Characteristics
Constitutional	Describes the molecular composition without using connectivity or structural geometry
Topological	Describes the atomic connectivity in the molecule
Information indexes	Based on calculation of the information content of a molecule
Electrostatic	Reflects the charge distribution
Geometrical	Describes molecular shape and volume on the basis of the 3D-coordinates of the atoms in a stable conformer
Quantum mechanical	Energy-related descriptors, based on quantum-mechanical calculations, which allows one to obtain the optimized chemical structure
Fragment-based descriptors	Describes the presence or absence of a variety of atom centred fragments, functional groups and fingerprint-type descriptors
Physicochemical properties	Includes several types of experimentally calculated or predicted (using specific software) physicochemical properties reflecting certain properties of the compound

In some cases the relationship is a simple rule: if a certain fragment is present, then the chemical is toxic. Formally, this approach is characteristic of SAR models and algorithms which search for the occurrence of such a fragment. In the case of quantitative models, the simplest possible equation is a linear one. For many years now the use of multiple parameters to generate more complex equations has been quite common. Figure 10.1 depicts schematically the steps involved in QSAR modelling, starting from the chemical structure, proceeding through calculations of descriptors and finally obtaining the predictive model.

Advanced mathematical techniques designed to extract knowledge from the data (also referred to as data mining), including artificial neural networks, genetic algorithms and fuzzy logic, are being used more and more commonly. Thus, on a conceptual basis, it is possible to distinguish between methods which codify knowledge that already exists (such as methods that focus on the occurrence of fragments known from previous examples to exert toxic effects) and attempts to extract new information implicitly present in large sets of data. Indeed, sometimes the terms explicit and implicit knowledge are applied in this context. We want to emphasize that even with so-called explicit knowledge, only a certain probability of an effect is obtained, both false positives and false negatives being entirely possible.

10.2.4 QSAR Models: Taking Three-dimensional Information into Consideration

The main difference between the 3D-QSAR and standard QSAR approach involves the type of structural information (chemical descriptors) employed to generate the predictive model. As indicated by its name, 3D-QSAR takes

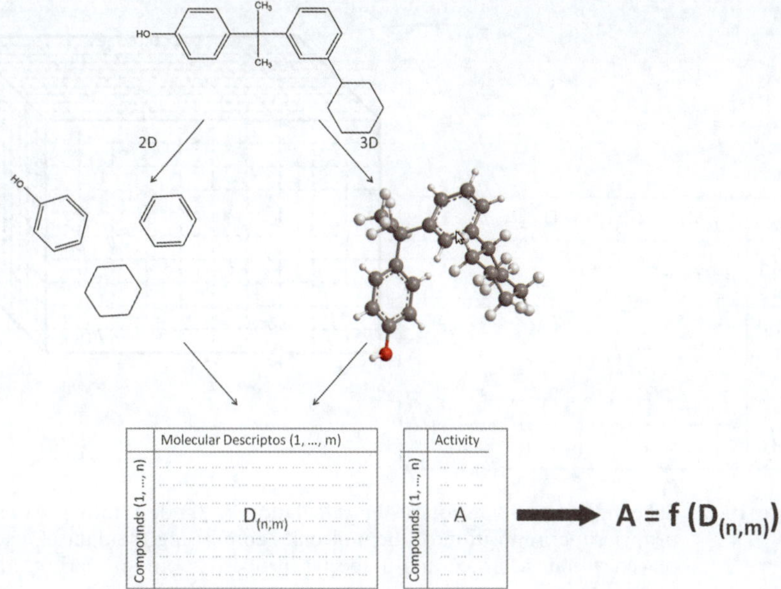

$$A = f\,(D_{(n,m)})$$

Figure 10.1 Schematic representation of the process of QSAR modelling. Several molecular descriptors (2D, 3D and fragments) are calculated from the chemical structure and used to compile the descriptor matrix (D). Each line of in this matrix represents a compound in the dataset characterized by the molecular descriptors (m) arranged in the columns. The activity matrix (A) depicts the activities of the chemicals in D.

into consideration information derived from the three-dimensional conformation of the compound, *i.e.* not only the properties derived from the connections between the atoms in the ligand, but also the properties determined by how these atoms are arranged in space, such as molecule surface, electrostatic interactions and geometric characteristics. This introduction of 3D information can help to explain structure–activity relationships of endocrine disruptors, since the newly developed models take into account energetic and geometrical interactions between important structural elements of the chemical and the binding site on the receptor. At the same time, inclusion of 3D structural information enhances both the complexity of the problem and the computational resources (time and computing power) required for simulation.

Extraction of molecular features from chemical compound's 3D structure for 3D-QSAR is accomplished most commonly at present employing the Molecular Interaction Fields (MIF)[6] and Comparative Molecular Field Analysis (CoMFA)[7] approaches. In both of these procedures the descriptors are calculated by placing the molecule of interest within a lattice and then utilizing a probe (*e.g.* an sp³ carbon atom), placed sequentially at each point of the grid, to calculate steric and electrostatic interactions (Figure 10.2). The field-based

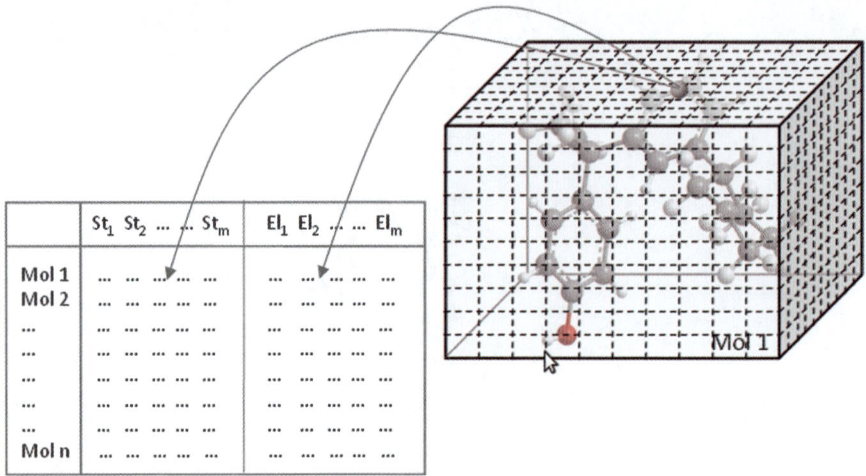

Figure 10.2   Schematic representation of the calculation of 3D descriptors. After the proper superimposition of the molecules comprising the dataset, each is placed within a lattice and a specific probe (*e.g.* an sp³ carbon atom) utilized to calculate the contributions by steric (St$_{1...m}$) and electrostatic (El$_{1...m}$) fields. All of these contributions are then employed as descriptors for QSAR modelling.

descriptors calculated in this manner and the resulting descriptor matrix can then be used in standard QSAR analysis.

The first step is to reduce the number of descriptors, since current methods supply thousands of these. Several statistical methods are applied to search for the smallest number of descriptors that can correlate the structures of the molecules in the dataset of interest to their activities. One of the most commonly used is the Partial Least Square (PLS), a regression analysis that can handle collinear variables and generate "latent variables", *i.e.* orthogonal linear combinations of the original set of descriptors. Another common approach to reducing the number of descriptors is Principal Component Analysis (PCA), by which a number of descriptors that may be correlated can be transformed into a smaller number of "new" and uncorrelated variables called *principal components*. This transformation is performed in such a manner that the first principal component exhibits the highest possible variance. Each following component must then be orthogonal to the previous one and also demonstrate the highest variance possible.

Another major difference between standard QSAR and 3D-QSAR is the importance of the correct 3D alignment of the molecules in the latter approach. This is essential for those molecules included in the datasets used to create and validate the model, as well as for applying the model to predict unknown activities of molecules. Such a correct alignment is necessary simply because the position of the molecule within the grid influences the field values (the descriptors).

The methods applied for correct alignment (superimposition) based on a common skeletal superimposition can be divided into two major classes:

1. Methods based on crystallographic (experimental) or docking (*in silico*) information and which begin with the position of the ligand inside the binding pocket.
2. Methods starting from a pharmacophoric hypothesis, a pharmacophore being defined by IUPAC as "an ensemble of steric and electronic features that is necessary to ensure the optimal supramolecular interactions with a specific biological target and to trigger (or block) its biological response".[8] The pharmacophoric features are arranged at particular positions in space and then used to align various molecules bearing all or some of these same features.

The modelling phase in 3D-QSAR provides, in addition to the model itself, a graphical representation of areas around the molecule that contribute (positively or negatively) to electrostatic and/or steric interactions that modulate the properties of the target. Such information cannot be obtained by standard QSAR, which only deals with the molecule as a whole and does not take local contributions by parts of the compound into consideration.

Other improvements associated with 3D-QSAR concern the more detailed description of the chemical structure obtained and the possibility of taking chirality (which is typically not evaluated in classical QSAR) into account. However, every coin has a flipside: 3D-QSAR models are typically applied to compounds with similar skeletal structures and can be influenced excessively by manual optimization of compounds. Accordingly, 3D-QSAR models require more extensive evaluation of their statistical robustness and ability to predict than do standard QSAR models.

10.2.5 The Docking Approach

Unlike the QSAR procedures, virtual docking includes information about the target molecule being screened.[9] This target is usually a protein (*i.e.* receptor), whereas the docking substance can be a macromolecule (such as DNA or another peptide) or a small molecule (ligand). In the context of food safety, and more precisely endocrine-disruptive activity, this approach is applied mainly when the target is a nuclear receptor, such as the estrogen (ER), androgen (AR) or aryl hydrocarbon receptor (AhR).[10]

Virtual docking requires the complete 3D structure of the receptor or at least of the portion (subunit) that contains the binding site, which is probably the most important and difficult step to achieve. The 3D structures of receptors (and of proteins in general) are obtained primarily by X-ray crystallography, NMR spectroscopy and/or homology modelling. Although all three methods obviously have their own pros and cons, X-ray crystallography is generally preferable since this procedure provides more precise positions for the atoms.

Homology modelling, on the other hand, is employed only when no structural information from one or both of the two other methods is available. This is because in this case the structure is obtained indirectly by aligning the sequence of the protein of interest to that of a homologous protein whose structure has already been determined.

Whatever the source of the structural information, several adjustments and refinements of this information are necessary for virtual docking. Since this procedure is designed to position a putative ligand into the binding pocket of its receptor, the conformation of the ligand must also be known, adding another complex variable to the situation.

With respect to both the binding site and the putative ligand, the first important question is: which conformers to use? A common approach is to perform repeated docking simulations with several stable conformers. A number of computational techniques can be employed to obtain different stable conformers of the same ligand or protein, including Molecular Mechanics (MM), Molecular Dynamics (MD), Monte Carlo sampling and *ab initio* calculations.

To simulate a docking of different conformers of a library of putative ligands to the binding site of a receptor, the position of the binding pocket must be defined. If the 3D structure of the receptor with a bound ligand has been determined by X-ray crystallography and/or NMR, most docking programs can utilize the known position of this ligand to position the potential ligands being screened. If insufficient information concerning the exact position of the binding site is available, several software tools (open source and commercial) can be used to predict binding pockets on protein surfaces.

Depending on the precision required, three types of simulated docking can be performed:

1. Rigid docking, in which the positions of all of the atoms in the protein binding site and the ligand are frozen and the rigid ligand is rotated inside the binding pocket. This approach results in the least detailed approximation of receptor–ligand interaction.
2. Flexible docking, in which all the atoms in the binding site and ligand are allowed to move freely, adjusting their position in response to interactions with surrounding atoms. This approach allows the most accurate approximation, but, even on powerful computers, can be very time consuming, especially when the ligand library to be screened is large.
3. Semi-flexible docking, which, as suggested by the name, is a compromise between the other two types of docking. In this case the atoms in the binding site are frozen, while those in the ligand are allowed to move freely.

For each putative ligand, several positions inside the binding pocket are evaluated with respect to stability in order to determine the most probable among all the possible protein–ligand complexes (exploration of the conformational space). The complementarity between the two molecules

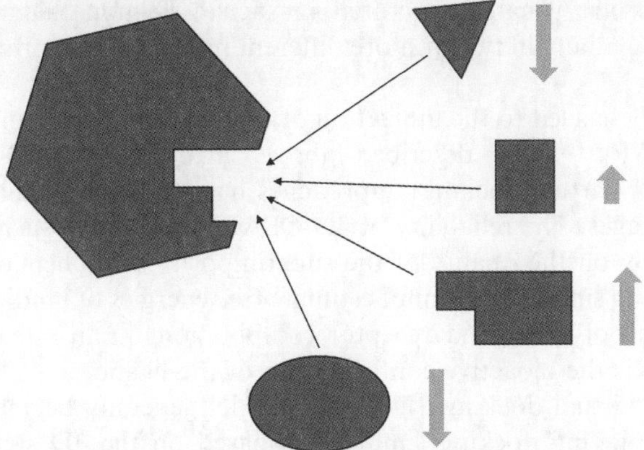

Figure 10.3 Evaluation of ligand–receptor complementarity. The precision of matching, based on steric and electrostatic interactions, provides a measure of the strength of the interaction between the receptor binding site and putative ligand. The vertical arrows indicate the binding strength.

(Figure 10.3) is scored on the basis of steric, electrostatic and/or other energetic interactions (*e.g.* the salvation effect) and the binding energy of the ligand–receptor complex is estimated.[11]

Virtual screening is widely used in the pharmaceutical industry as an early step in drug discovery. In this connection, docking allows exploration of large libraries of putative ligands in order to identify those that are physically and energetically capable of binding to the target receptor. However, the scoring used in virtual screening cannot estimate the binding energy accurately enough to reflect that obtained experimentally *in vivo*.

Such screening by virtual docking should be and, in fact, has been applied successfully to the risk evaluation of food contaminants, including endocrine disruptors.[12–14] Virtual docking is often used in combination with other *in silico* techniques to enhance the reliability of predictions. As in the case of drug design, the primary use of this approach is to pre-screen and prioritize food chemicals for the subsequent experimental procedures. Virtual screening is also being adopted by EC-funded projects that focus on food safety. For example, the CASCADE project (http://www.cascadenet.org/) has utilized docking simulation to successfully screen chemical contaminants in food and their metabolites.[15,16]

10.2.6 Combining Docking and 3D-QSAR to Improve Predictive Power

A common strategy is to apply several different modelling techniques for the same endpoint, in order to extract as much information as possible from datasets and thereby improve predictive power.[17–19] These so-called hybrid models are usually structured in such a fashion that a chemical is evaluated sequentially in two or more different models until a prediction can be made.

This strategy has led to the utilization of docking simulation in combination with 3D-QSAR.[20–23] As described above, ligand-based (3D-QSAR) and receptor-based (virtual docking) approaches have their individual advantages and disadvantages. The reliability of 3D-QSAR predictions (such as CoMFA) depends strictly on the quality of the superimposition of chemical structures, whereas docking simulation cannot estimate free energies of binding sufficiently well to rank all of the ligand–receptor combinations in an attempt to decide *a priori* which is the bioactive conformation of the ligand.

The hybrid virtual docking 3D-QSAR model generally begins with virtual screening, employing docking simulations based on the 3D structure of the receptor–ligand complex as determined by X-ray crystallography. Simulation by techniques of molecular mechanics (such as simulated annealing) is subsequently employed to compute the most stable conformation of the co-crystallized ligand and the position thus obtained used as a reference for the superimposition of the chemicals being screened. Starting in this way, from a more accurate superimposition of the chemicals, results in a more accurate and reproducible calculation of the field-based descriptors and thereby in a 3D-QSAR model with improved predictive power.

10.3 One Example of an Integrated *In Silico* Approach to the Evaluation of Endocrine-disrupting Activity Mediated Through the Estrogen Receptor

The best use of *in silico* procedures for prioritizing or assessing endocrine-disrupting potential is obtained by integrating a number of the various approaches discussed above, some of which provide more detailed information than others. On the other hand, computational time is directly proportional to the level of precision attained. Accordingly, the pharmaceutical industry utilizes *in silico* approaches for a rapid screen of large libraries of compounds to identify candidate molecules (*e.g.* chemicals that might possess the desired pharmacological activity) for subsequent evaluation with more sophisticated tools and experimental investigation.

A similar approach has been applied by the FDA's National Center for Toxicological Research (NCTR) to the study of effects mediated by the estrogen receptor (ER),[24] which is among the nuclear receptors involved in endocrine disruption that have been investigated most extensively. The NCTR conducted their study in four phases arranged hierarchically to reduce the size of the compound, enhance the precision of prediction and minimize the number of false negatives. Within each phase, different models were employed in a complementary manner.

In more in detail, these four phases were as follows:

- Phase 1: a "filtering" step in which very simple parameters such as molecular weight and the presence or absence of ring structures were considered.
- Phase 2: a classification process designed to discriminate between ligands that bind to the ER and those that do not. This process involved a total of 11 models based on different paradigms and all chemicals evaluated as positive in at least one model were passed on to the third phase. The 11 models in this phase are three structural alerts (structural fragments associated with ER binding), seven pharmacophores (a combination of several 3D molecular features required to exhibit activity) and a decision tree (*i.e.* a QSAR model based on classification trees).
- Phase 3: quantification of the strength of interaction of the compounds with the ER, utilizing a CoMFA model.
- Phase 4: the final step in which other factors were taken into account to arrive at a final list of compounds of concern that might require further evaluation. This step incorporates human knowledge (in terms of rules), as well as consideration of other issues relevant to the determination of priorities.[25]

10.4 Conclusions

We describe here a wide variety of *in silico* tools that can be applied to the evaluation of, for example, dietary endocrine disruptors. In some cases, adverse effects can be predicted simply on the basis of the compound's chemical structure, applying the QSAR strategy to obtain information related to the presence of certain moieties or more general global descriptors. Indeed, several endocrine-disrupting substances contain certain specific fragments, such as a phenol groups, which thus can be employed to identify probable endocrine disruptors.

Other programs analyze the chemical structure of a compound on the basis of global descriptors. With such approaches, it is easier to quantify potency, which can be highly useful in identifying chemicals with a phenolic ring that are not very active, such as many natural components of food containing a steroid structure. These procedures can be utilized to model complex, multi-factorial phenomena, *i.e.* not simply binding to a nuclear receptor, but also more complex, multi-step processes.

The so-called 3D-QSAR strategy is better suited to the exploration of sets of chemicals that are closely related from a structural point of view. In some ways, this represent a limitation in comparison to previous QSAR models applicable to more heterogeneous sets of compounds. At the same time, 3D-QSAR can provide more detail on a certain portion of a given substance that may play a role in the adverse phenomenon, as well as being able to handle and distinguish between chiral compounds, which classical QSAR models cannot.

Virtual docking provides much more detailed information at the mechanistic level, *i.e.* information concerning binding to the nuclear receptor. Several different receptors can be tested if their structures are known. Such structural determination is both a requirement and an opportunity, since the information obtained is then very detailed and directly related to binding. One obvious limitation here is that *in vivo* effects may involve many processes other than binding to the receptor, *e.g.* the production of a metabolite with a different binding affinity or modulation in some way of the overall effect initiated by binding.

Some *in silico* procedures can easily deal with very large sets of compounds, whereas others are limited to homogeneous sub-sets even requiring that each chemical be evaluated manually one at a time. Another important aspect is that some of these methods are quite simple, requiring almost no training, whereas others are much more complicated. The price is also highly variable, some being free and others quite expensive.

Owing to these technical characteristics and, of course, their theoretical basis, certain procedures are more useful for rapid screening of contaminants than others. Ease of performance and low cost also promote broad assessment of the natural components of food, *e.g.* evaluation of the carcinogenic potential of such components *in silico*.[26] Methods such as 3D-QSAR and virtual screening are more suitable for a refined examination of effects, even though these procedures have also been used to screen the binding of hundreds of chemicals to nuclear receptors by virtual docking, for instance, within the EC-funded CASCADE and CHEMOMENTUM projects.

We propose that combining these different approaches may open new perspectives for the future. *In silico* methods are usually considered as alternatives to animal models, meant to mimic the *in vivo* situation as closely as possible, but other perspectives are also possible. Novel experimental models for endocrine-disrupting effects are clearly needed and, in this connection, *in silico* approaches can be valuable supplementary tools, especially in combination with other data.

References

1. E. Lo Piparo and A. Worth, *Review of QSAR Models and Software Tools for Predicting Developmental and Reproductive Toxicity*, Publications Office of the European Union, Luxembourg, 2010.
2. T. W. Schultz, M. T. D. Cronin, J. D. Walker and A. O. Aptula, *J. Mol. Struct.*, 2003, **622**, 1.
3. R. Kroes, A. G. Renwick, M. Cheeseman, J. Kleiner, I. Mangelsdorf, A. Piersma, B. Schilter, J. Schlatter, F. van Schothorst, J. G. Vos and G Würtzen, *Food Chem. Toxicol.*, 2004, **42**, 65.
4. S. J. Enoch, in *Recent Advances in QSAR Studies – Methods and Applications*, ed. T. Puzyn, J. Leszczynski and M. T. D. Cronin, Springer, Dordrecht, 2010, p. 209.
5. R. Todeschini and V. Consonni, *Handbook of Molecular Descriptors*, Wiley-VCH, Weinheim, 2000.

6. G. Cruciani, *Molecular Interaction Fields: Applications in Drug Discovery and ADME Prediction*, Wiley-VCH, Weinheim, 2006.
7. R. D. Cramer, D. E. Patterson and J. D. Bunce, *J. Am. Chem. Soc.*, 1988, **110**, 5959.
8. C. G. Wermuth, C. R. Ganellin, P. Lindberg and L. A. Mitscher, *Pure Appl. Chem.*, 1998, **70**, 1129.
9. W. P. Walters, M. T. Stahl and M. A. Murcko, *Drug Discovery Today*, 1998, **7**, 903.
10. M. A. Lill and A. Vedani, in *Computational Toxicology – Risk Assessment for Pharmaceutical and Environmental Chemicals*, ed. S. Ekins, Wiley-Interscience, New York, 2007, p. 315.
11. T. I. Oprea and R. Garland, *Perspect. Drug Discovery Des.*, 1998, **9**, 35.
12. B. Wu, T. Ford, J. D. Gu, X. X. Zhang, A. M. Li and S. P. Cheng, *Chemosphere*, 2010, **80**, 535.
13. T. Nose, T. Tokunaga and Y. Shimohigashi, *Toxicol. Lett.*, 2009, **191**, 33.
14. W. Yang, Y. Mu, J. P. Giesy, A. Zhang and H. Yu, *Chemosphere*, 2009, **75**, 1159.
15. E. Lo Piparo, K. Koehler, A. Chana and E. Benfenati, *J. Med. Chem.*, 2006, **49**, 5702.
16. G. Lemaire, C. Benod, V. Nahoum, A. Pillon, A. M. Boussioux, J. F. Guichou, G. Subra, J. M. Pascussi, W. Bourguet, A. Chavanieu and P. Balaguer, *Mol. Pharmacol.*, 2007, **72**, 572.
17. A. Lombardo, A. Roncaglioni, E. Boriani, C. Milan and E. Benfenati, *Chem. Central J.*, 2010, **4**, S1.
18. T. Ferrari and G. Gini, *Chem. Central J.*, 2010, **4**, S2.
19. E. Benfenati, *Quantitative Structure–Activity Relationship (QSAR) for Pesticide Regulatory Purposes*, Elsevier, Amsterdam, 2007.
20. W. Sippl, *J. Comput. Aided Mol. Des.*, 2000, **14**, 559.
21. W. Sippl and H. D. Höltje, *J. Mol. Struct.*, 2000, **503**, 31.
22. W. Sippl, *Bioorg. Med. Chem.*, 2002, **10**, 3741.
23. W. Sippl, *J. Comput. Aided Mol. Des.*, 2002, **16**, 825.
24. L. Shi, W. Tong, H. Fang, Q. Xie, H. Hong, R. Perkins, J. Wu, M. Tu, R. M. Blair, W. S. Branham, C. Waller, J. Walker and D. M. Sheehan, *SAR QSAR Environ. Res.*, 2002, **13**, 69.
25. W. Tong, H. Fang, H. Hong, Q. Xie, R. Perkins and D. M. Sheehan, in *Predicting Chemical Toxicity and Fate*, ed. M. T. D. Cronin and D. Livingstone, CRC Press, Boca Raton, 2004, p. 285.
26. L. G. Valerio Jr., K. B. Arvidson, R. F. Chanderbhan and J. F. Contrera, *Toxicol. Appl. Pharmacol.*, 2007, **222**, 1.

CHAPTER 11

Application of Percellome Toxicogenomics to Food Safety

J. KANNO,[a] K. AISAKI,[a] K. IGARASHI,[a] N. NAKATSU,[a,b] Y. KODAMA,[a] K. SEKITA,[a] A. TAKAGI[a] AND S. KITAJIMA[a]

[a] National Institute of Health Sciences, Tokyo 158-8501, Japan; [b] National Institute of Biomedical Innovation, Osaka 567-0085, Japan

11.1 Introduction

The approaches employed to assess chemical toxicity are seldom applied to food safety assessment for a number of reasons. In the first place, routine toxicity assessment involves administration of a pure compound to experimental animals at different doses up to the maximal tolerated dose (MTD) to determine the no observed (adverse) effect level [NO(A)EL], while food is a complex mixture of chemicals. Secondly, the sensitivity of routine toxicity tests is inadequate for foods. The maximal amount of the test substance applied is limited by guidelines to 5% of the diet or, in the case of oral gavage, $10\ \mathrm{mL\ kg^{-1}}$ (approximately $10\ \mathrm{g\ kg^{-1}}$ for liquids), in order not to jeopardize reliability (especially with respect to chronic toxicity and carcinogenesis) by inducing malnutrition. For example, low body weight alone can result in tumors at low incidences.[1–3] When an item of food is found to be non-toxic at the dose of $10\ \mathrm{g}$ $\mathrm{kg^{-1}}$, this value is taken as NOEL and the tolerable daily intake (TDI) calculated, using a safety factor of 1/100, to be $100\ \mathrm{mg\ kg^{-1}}$. Thus the grams or even tens of grams per kg body weight that we consume daily are not allowed in routine toxicity testing and, moreover, according to such standard approaches, if a food showed any toxicity in routine testing, such food would be highly toxic.

Issues in Toxicology No. 11
Hormone-Disruptive Chemical Contaminants in Food
Edited by Ingemar Pongratz and Linda Vikström Bergander
© Royal Society of Chemistry 2012
Published by the Royal Society of Chemistry, www.rsc.org

Even though flavorings are added to food and other consumed products at relatively low concentrations, these substances are often not subject to toxicity testing at the MTD from other reasons. One practical consideration is that the odor of many flavorings at MTD is too strong to be tolerated by the test animal or the staff that care for them. Moreover, in some cases, the amount of a flavoring required to conduct an MTD test is more than its annual production and/or prohibitively costly. At the practical dose levels, flavoring does not exert any detectable histopathological effect on experimental animals. However, such doses do stimulate, at least, olfactory (and perhaps other) receptors. Thus, more sensitive approaches to assessing the toxicity of food and flavors are highly desirable. One approach is to use particularly sensitive animal models, such as transgenic mice to evaluate certain biological phenomena, including carcinogenic potential. Another strategy is to detect early molecular changes that occur prior to morphological alterations and/or other symptoms. Toxicogenomics is one example of the latter.

Our Percellome Toxicogenomics Project is designed to identify dynamic and extensive networks of genes whose time- and dose-dependent patterns of expression in response to a chemical, or a mixture of chemicals such as food, allows its toxic effects to be predicted. Here, we describe this project briefly, employing a flavoring compound, estragole, as a peroxisome proliferator-activated receptor alpha (PPAR-alpha) agonist whose effects are similar to those of clofibrate. Among the more than 100 chemicals tested, 45 are food related (Table 11.1), including food additives, components of food, functional health foods, agrochemicals, and other food contaminants (*e.g.* that might leach out of packaging materials).

11.2   Materials and Methods

Forty-eight male C57BL/6 Cr Slc or C57BL/6J mice (SLC, Hamamatsu, Charles River Japan, Atsugi, Japan) were maintained for two weeks in an enclosed environment (diet: CRF-1, Oriental Yeast Co.) with a 12-h photoperiod (starting at 8:00 a.m.) and then, at 12 weeks of age, a single oral administration of the test chemical at 10:00 ~ 10:20 a.m. and at the doses indicated in Table 11.1. At 2, 4, 8, and 24 hours later, the livers of three animals at each dose were collected. To minimize circadian oscillations in the transcriptome, 12 animals were processed at each time-point at a rate of 2.5 min from initiation ether anesthesia to immersion of the liver and other organ samples into RNAlater solution (Ambion, TX, USA), thereby keeping the total elapsed to within 30 min. Prior to analysis, the blocks of liver tissue were soaked in RNAlater overnight at 4 °C. This solution was then replaced with RLT buffer (Qiagen, Germany), the tissue homogenized, and a small aliquot of each homogenate treated with DNAse-free RNase A, followed by proteinase K; the DNA concentration was then determined with PicoGreen fluorescent dye (Molecular Probes, USA). Next, on the basis of its DNA content, each sample was spiked with the same relative amount of GSC cocktail (containing

Table 11.1 Food-related chemicals in Percellome Database.

Food-related Chemicals (including agrochemicals)	CAS No.	Dose (mg/kg)	Vehicle	Percellome study No. TTG
Acephate	30560-19-1	0, 0.7, 2, 7, 20, 70 mg/kg	MC 0.5%	67, 109
Alpha lipoic acid	1077-28-7	0, 10, 30, 100 mg/kg	MC 0.5%	128
Aluminum ammonium sulfate	7784-26-1	0, 30, 100, 300 mg/kg	MC 0.5%(WAKO)	155
Aluminum lactate	18917-91-4	0, 30, 100, 300 mg/kg	MC 0.5%(WAKO)	163
Aluminum sulfate	10043-01-3	0, 100, 300, 1000 mg/kg	MC 0.5%(WAKO)	178
Bisphenol A	1980-5-7	0, 30, 100, 300 mg/kg	DMSO 0.1% + MC 0.5%	47
Caffeine	1958-8-2	0, 3, 10, 30 mg/kg	MC 0.5%	59
Carbaryl	63-25-2	0, 2, 7, 10, 20, 30, 100 mg/kg	MC 0.5%	68, 166
Chlorpyrifos	2921-88-2	0, 3, 10, 30 mg/kg	MC 0.5%(WAKO)	165
Citric acid-calcium salt	813-94-5	0, 100, 300, 1000 mg/kg	MC 0.5%	39
Coenzyme Q10	303-98-0	0, 30, 100, 300 mg/kg	corn oil	86
Curcumin	458-37-7	0, 70, 200, 700 mg/kg	corn oil	132
DEHP (di(2-ethylhexyl)phthalate)	117-81-7	0, 200, 700, 2000 mg/kg (*cf.* MEHP)	corn oil	98
Dexamethasone	1950-2-2	0, 10, 30, 100 ug/kg	DMSO 0.1% + MC 0.5%	62
Estragole	140-67-0	0, 10, 30, 100 mg/kg	corn oil	147
Ethanol	64-17-5	0, 150, 500, 1500 mg/kg (d = 0.79)	MC 0.5%	78
Food Red No.104	18472-87-2	0, 10, 30, 100 mg/kg	MC 0.5%(WAKO)	156
Forskolin	66575-29-9	0, 1, 3, 10 mg/kg	MC 0.5%	60
Genistein	446-72-0	0, 10, 30, 100 mg/kg	DMSO 0.1% + MC 0.5%	48
Glycyrrhizin2K	1405-86-3	0, 3, 10, 30 mg/kg	DMSO 0.1% + MC 0.5%	110
Hydroxycitric Acid	27750-10-3	0, 100, 300, 1000 mg/kg	MC 0.5%	38

Indigo	482-89-3	0, 30, 100, 300 mg/kg	corn oil	57
Maltol	118-71-8	0, 10, 30, 100 mg/kg	corn oil	133
MEHP (mono(2-ethylhexyl)phthalate)	4376-20-9	0, 70, 200, 700 mg/kg (*cf*. MEHP)	corn oil	104
Methanol	67-56-1	0, 30, 100, 300 mg/kg (*cf*. Formalin, EtOH)	MC 0.5%	79
Methoprene	40596-69-8	0, 0.3, 1, 3 mg/kg	DMSO 0.1% + corn oil	66
Methyl dihydro jasmonate	24851-98-7	0, 10, 30, 100 mg/kg	corn oil	135
Nerolidol	cis: [3790-78-1], trans:[40716-66-3]	0, 5.9, 17.7, 59.1 mg/kg	corn oil	134
Paraquat dichloride	1910-42-5	0, 1, 3, 10 mg/kg	MC 0.5%	61
Pentachlorophenol	87-86-5	0, 10, 30, 100 mg/kg	MC 0.5%	16
Permethrin	52645-53-1	0, 10, 30, 100 mg/kg	corn oil	97
Phytol	7541-49-3	0, 10, 30, 100 mg/kg	corn oil	136
Pyriproxyfen	95737-68-1	0, 0.3, 1, 3 mg/kg	DMSO 0.1% + corn oil	87
Sodium Dehydroacetate	4418-26-2	0, 30, 100, 300 mg/kg	MC 0.5%(WAKO)	154
Tebufenozide	112410-23-8	0, 0.3, 1, 3 mg/kg	DMSO 0.1% + corn oil	88
Tributyltin	56573-85-4	0, 30,100, 300 ug/kg	DMSO 0.5% + MC 0.5%	85
Tributyltin chloride	1461-22-9	0, 3,10, 30 ug/kg	DMSO 0.1% + MC 0.5%	49
Verbenone	18308-32-5	0, 10, 30, 100 mg/kg	corn oil	148
Warfarin (sodium salt)	81-81-2	0, 10, 30, 100 mg/kg	MC 0.5%	81
FD&C Blue No.1 (brilliant blue)	3844-45-9	0, 30, 100, 300 mg/kg	MC 0.5%(WAKO)	152
FD&C Blue No.2 (Indigo carmine)	860-22-0	0, 30, 100, 300 mg/kg	MC 0.5%(WAKO)	139
Food Red No.102	2611-82-7	0, 100, 300, 1000 mg/kg	MC 0.5%(WAKO)	177
FD&C Red No.40	25956-17-6	0, 30, 100, 300 mg/kg	MC 0.5%(WAKO)	140
FD&C Green No.8 (pyranine conc)	6358-69-6	0, 30, 100, 300 mg/kg	MC 0.5%(WAKO)	176
FD&C Yellow No.5 (Food Yellow No.4 Japan)	1934-21-0	0, 100, 300, 1000 mg/kg	MC 0.5%(WAKO)	161

MC: methylcellulose

five *Bacillus subtilis* RNA sequences corresponding to the Affymetrix GeneChip probe sets AFFX-ThrX-3_at, AFFX-LysX-3_at, AFFX-PheX-3_at, AFFX-DapX-3_at, and AFFX-TrpnX-3_at), and subsequently processed in accordance with the Affymetrix Standard protocol for application to Mouse 430 2.0 GeneChips. After conversion of the raw read-outs to Percellome data,[4] the values obtained (with four doses and five time-points, using 2-h vehicle value as 0-h) were plotted on the three-dimensional graphs (*cf.* http://toxicomics. nihs.go.jp/db/) (Figure 11.1).

The highest dose that, 24 hours after a single oral gavage, did not influence body weight or cause macroscopic changes in internal organs, microscopic alterations in liver observed with H&E staining, and/or abnormal behavior was employed as the maximal dose. In general, the maximal doses in the Percellome studies are 10- to 100-fold lower than the reported MTDs.

The Percellome data set was first screened with an "RSort" program,[5] which identifies the number of peaks on a surface, distinguishes between the surfaces for 45 101 probe sets simply on the basis of their shapes, includes which of the peaks is significantly different ($p < 0.05$) from the corresponding control (vehicle) values, as evaluated by Student's *t*-test, and excludes probe sets with less than one copy per liver cell. Thereafter, taking into consideration the *t*-test values and ± 1 standard deviation, the surface shape was checked manually by

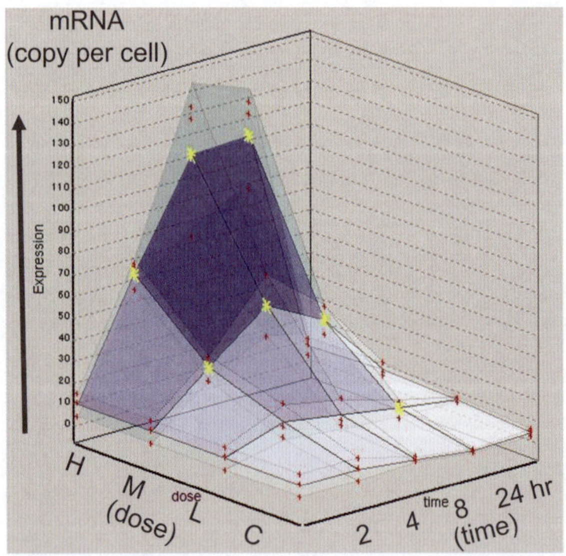

Figure 11.1   A three-dimensional representation of Percellome data. The dose- and time-dependent alterations in the expression of a probe set of genes is expressed as the absolute number of mRNA molecule per cell on the vertical axis. The mean of data from three different animals at each time and dose is represented, along with the SD (colored more lightly). Asterisks indicate $p < 0.05$ from concurrent control by student's *t*-test; small crosses represents data for each animal.

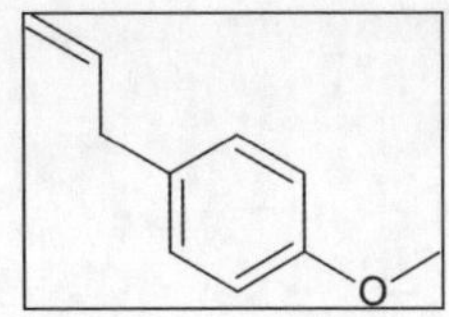

Figure 11.2 The chemical structure of estragole.

S.K. and/or J.K. Another original program referred to as "Percellome Explorer"[5] was utilized to compare pairwise the RSort results of all evaluations in the Percellome project with lists of shared probe sets.

The flavor compound estragole (1-allyl-4-methoxybenzene; CAS no. 140-67-0) (Figure 11.2) is a major component of tarragon, as well as of oils obtained from basil and other plants. We found that pure estragole (98%) from Sigma-Aldrich (catalog no. A29208-25G, lot no. 05202AH) at a maximal dose of 100 mg kg^{-1} caused no observable effects after 24 hours, so the Percellome investigation was conducted at doses of 0, 10, 30, and 100 mg kg^{-1} dissolved in corn oil (Sigma-Aldrich, C8267) and delivered by gavage.

11.3 Results

First, the Percellome data obtained were filtered automatically with the RSort program to identify surfaces with less than four peaks in the 4 × 4 matrix, a p-value of <0.05 at any time-point when comparing the control value to the corresponding maximal or middle dose, and a maximal peak of more than three copies of mRNA per cell. In this manner, 1214 up-regulated and 244 down-regulated sets were automatically selected out of the 45 101 probe sets in the mouse GeneChip, and subsequently 167 of those were selected manually as biologically feasible. The surfaces of several of the probe sets selected are depicted in Figure 11.3. Among those up-regulated, Ingenuity Pathway Analysis (IPA) highlighted a certain number genes encoding proteins associated with the PPAR-alpha. Of the metabolic enzymes up-regulated, Cyp4a10, Cyp4a14, and Cyp4a31, the last has been reported to be regulated by PPAR-alpha. Expression of the up-regulated transporters Slc16a5, Slc25a20, and Slc27a1 is also considered to be regulated by PPAR-alpha. A limited number of genes related to oxidative stress, apoptosis, and cell proliferation were induced (Table 11.2). The genes down-regulated could not be assigned to any known pathway by IPA or by searching the literature. Therefore, we conclude that administration of a single gavage of estragole up-regulates the expression of genes controlled by PPAR-alpha.

11.4 Discussion

Among others, PPAR-alpha regulates a set of genes whose products participate in the β-oxidation of fatty acids, as well as, in rodents at least, peroxisome

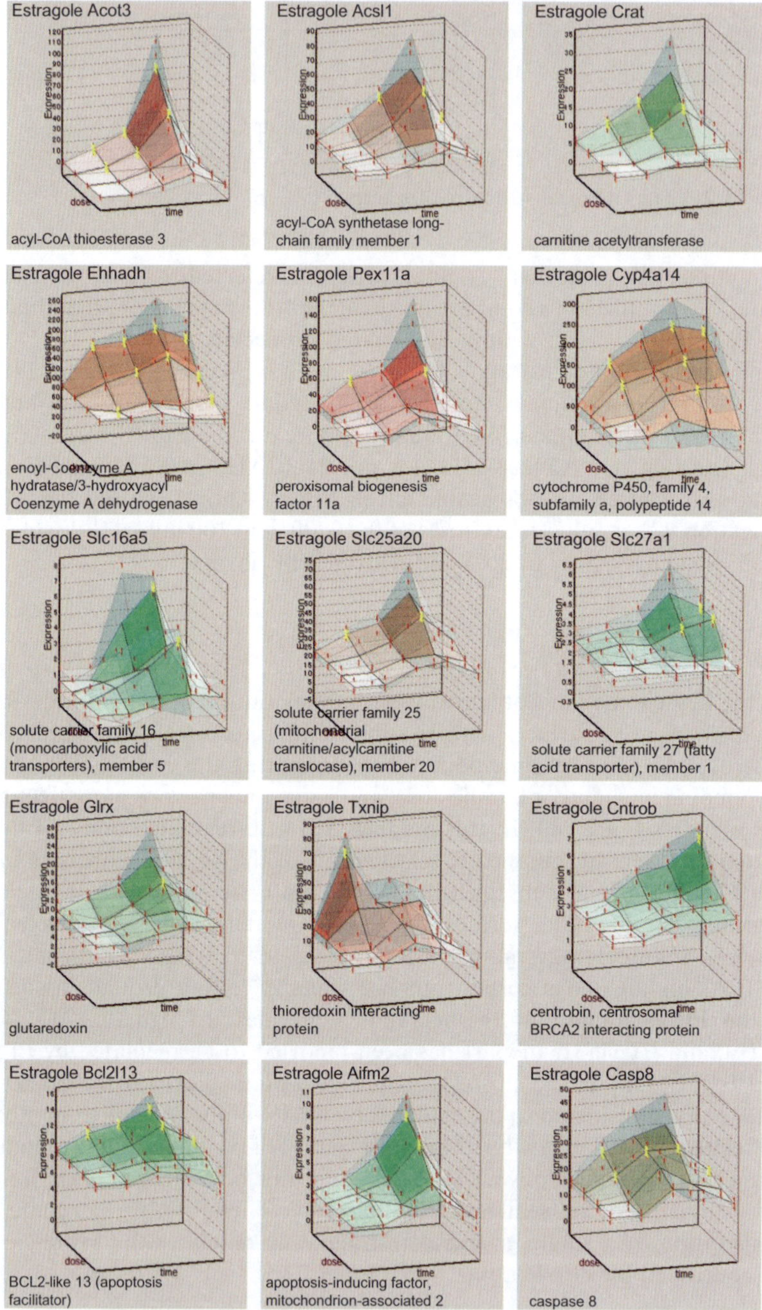

Figure 11.3 Representative three-dimensional surfaces of probe sets induced by estragole; Acot3, Acsl1, Crat, Ehhadh, Pex11a, Cyp4a14, Slc16a5, Slc25a20, and Slc27a1 are considered to be regulated by PPAR-alpha; the findings with respect to genes related to oxidative stress, apoptosis, and cell proliferation are also shown (*cf*. Table 11.2).

Table 11.2 Selected list of probe sets induced by estragole.

PPAR alpha pathway by IPA	
Acaa1a	1456011_x_at
Acot1 /// Acot2 /// LOC100044830	1422997_s_at
Acot1 /// LOC100044830	1449065_at
Acot3	1422925_s_at
Acot4	1422076_at, 1422077_at
Acox1	1444518_at
Acsl1	1460316_at
Acsl5	1428082_at
Agpat6	1422841_at, 1450776_at
Crat	1417008_at
Ehhadh	1448382_at
Pdk4	1417273_at
Pex1	1428716_at
Pex11a	1419365_at, 1449442_at
Pxmp4	1455438_at
P450s	
Cyp4a10 /// Cyp4a31	1424853_s_at
Cyp4a14	1423257_at
Cyp4a31	1440134_at
Transporters	
Slc12a7	1418257_at
Slc14a2	1426109_a_at
Slc16a1	1415802_at
Slc16a5	1434473_at
Slc22a5	1421848_at
Slc22a5	1450395_at
Slc23a2	1417329_at
Slc23a2	1417330_at
Slc25a20	1423108_at
Slc25a20	1423109_s_at
Slc25a42	1424790_at
Slc27a1	1422811_at
Slc29a3	1455731_at
Oxidative stress, apoptosis and cell proliferation	
Aifm2	1431143_x_at
Bcl2l13	1429539_at
Casp8	1424552_at
Cntrob	1433958_at
Glrx	1416592_at
Txnip	1415996_at
Others	
Dnaic1	1437093_at, 1437094_x_at
Fabp2	1418438_at
Gyk	1422703_at, 1422704_at
Klf10	1416029_at
Klf11	1437241_at
Oplah	1424359_at
Paqr7	1435312_at, 1460674_at

proliferation. Agonists of this receptor are carcinogenic towards rodent liver *via* mechanisms related to peroxisome proliferation.[6] At the same time, since peroxisome proliferation leads to lowering of serum triglyceride levels, PPAR-alpha is a molecular target for hypolipidemic drugs, such as clofibrate.

Here, we found that estragole induces a considerable number of genes known to be regulated by PPAR-alpha. The Percellome Explorer (PE) program, which automatically compares probe set lists, has identified clofibrate (81 probe sets), followed by di(2-ethylhexyl) phthalate (DEHP) (126 probe sets) as the two chemicals causing changes most similar to those evoked by estragole upon testing with the same experimental protocol (Figure 11.4) (the absolute numbers of common probe sets were normalized to the product of the numbers of altered probe sets in each study). The PE program also compares probe sets on the basis of time, in the present case 2 hours after administration. Such early-responding genes are considered to be direct or almost direct targets of a receptor(s) that is activated by the test chemicals. Again, the similarity of its pattern to those of clofibrate and DEHP[7,8] confirms that one of the primary receptors that binds estragole is the PPAR-alpha. Isozyme 4 of pyruvate dehydrogenase kinase (Pdk4), shown in Figure 11.4, has already been reported to be regulated by this latter compound. Ingenuity pathway analysis was then applied for comparison with reported information (Figure 11.5). Two hours after administration, TXNIP, Klf10, and KLF11 did not appear to be directly regulated by PPAR-alpha, even though these three genes have recently been reported to be under the regulation of this receptor,[9,10] albeit with no clear relationship to liver function. Dynein, axonemal, and intermediate chain 1 (Dnaic1) is induced 8 and 24 hours after administration of estragole and clofibrate, respectively, as well as by DEHP and its metabolite mono(2-ethylhexyl) phthalate (MEHP) at later time-points (Figure 11.6). Dnaic1 appears to play an important role in peroxisome biogenesis.[11,12] This delayed induction, compared to that of early-response genes such as Pdk4, might indicate that several mediators are located between PPAR-alpha and this gene. The nature of the pathway, together with a literature search, indicate that HNF4A and PEX13[13] may be involved, since the latter is induced slightly by estragole (not significantly) and clearly by clofibrate, DEHP, and MEHP (Figure 11.6).

Among the many isozymes of cytochrome P-450, Cyp4a10, Cyp4a14, and Cyp4a31 are induced significantly by estragole, as well as by clofibrate and DEHP. On the other hand, Cyp1a2, which is induced slightly by phenobarbital, was not up-regulated by estragole. All three of these Cyp4a's are considered to be regulated by PPAR-alpha. Whereas DEHP also induces Cyp2b10 and Cyp51, estragole and clofibrate do not. Phenobarbital also induces Cyp2b10 and Cyp51, but not the Cyp4a's. Thus, the present Percellome analysis indicates (Figure 11.7) that DEHP activates at least two receptors, both PPAR-alpha and the constitutive androstane receptor (NR1I3 or CAR), which regulates Cyp2b[14] and probably Cyp51.[15,16] In contrast, estragole and clofibrate activate PPAR-alpha, but not CAR.

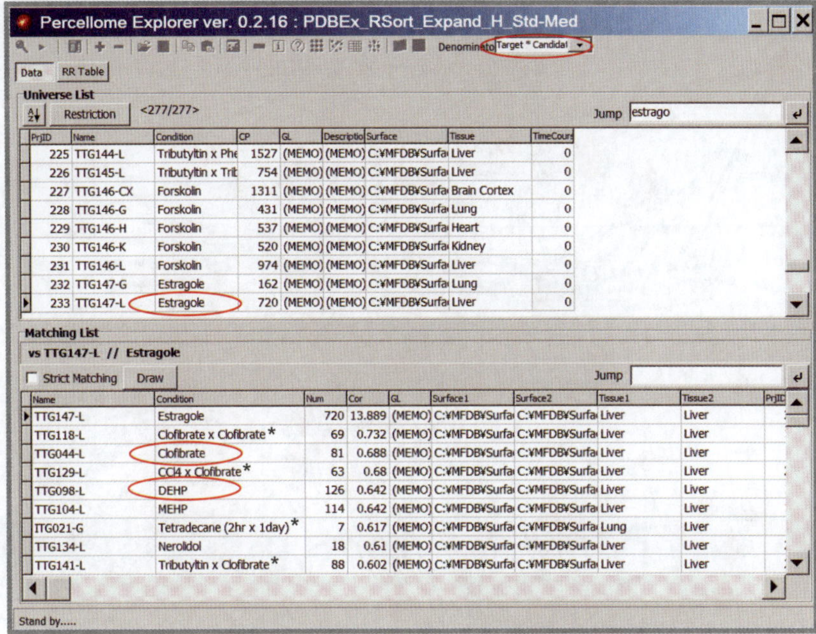

*:Performed by different protocols

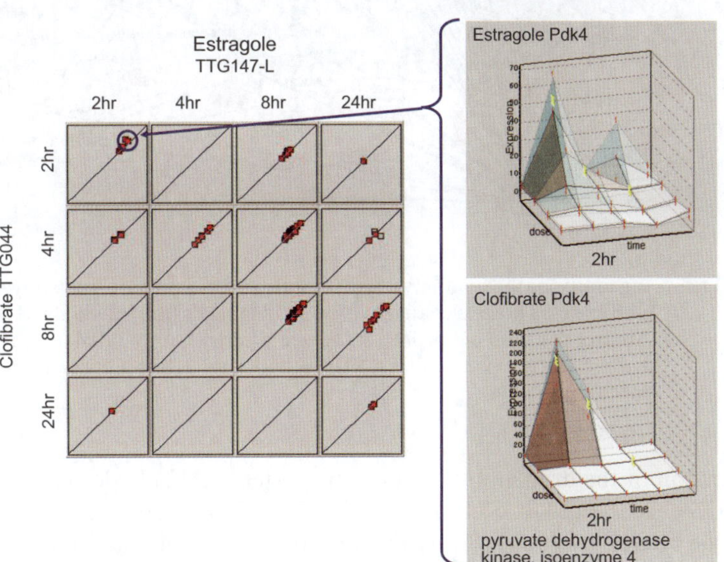

Figure 11.4 Comparison of the similarity between various Percellome studies utilizing the Percellome Explorer program. The upper figure shows the calculations that identify clofibrate and DEHP as the top chemicals most similar to estragole. The lower figure illustrates the similar peak times of the probe sets common to estragole and clofibrate, using Pdk4 (1417273_at) as an example.

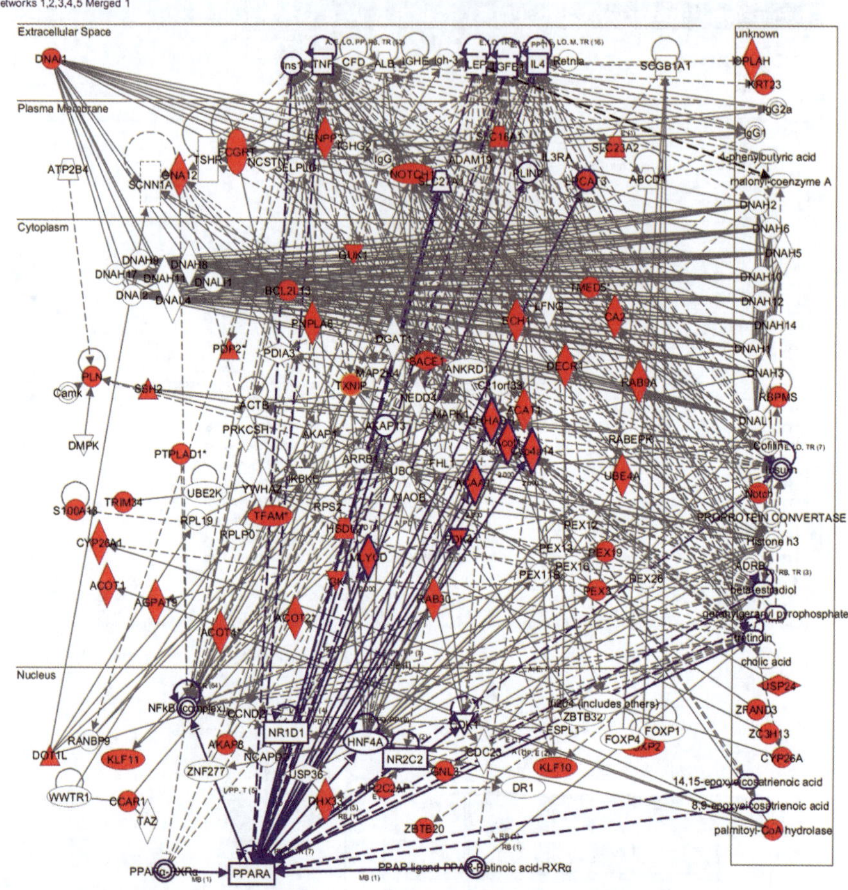

Figure 11.5   Ingenuity pathway analysis of the probe sets common to estragole and
clofibrate. Those elevated by both chemicals are highlighted in red.
Among the four 2-h probe sets (green asterisk), only Pdk4 appears to be
a direct target, whereas TXNIP, Klf10, and Klf11 appear to be influenced indirectly.

Expression of the drug and/or metal transporters Slc12a7, Slc14a2, Slc16a1,
Slc16a5, Slc22a5, Slc23a2, Slc25a20, Slc25a42, Slc27a1, and Slc29a3 appeared
here to be altered by estragole. Again, PE analysis reveals that most of these
same Slc transporters are also induced by DEHP and clofibrate (not shown).
Among these, only Slc27a1 has so far been reported to be regulated by PPAR-
alpha.[17,18] However, expression of most of these Slc transporters, including
Slc27a1, peaked after 8 hours, indicating that these genes may be activated
indirectly.

Our novel finding that estragole activates PPAR-alpha signaling may help
elucidate the mechanism(s) underlying estragole-induced carcinogenesis, *i.e.*

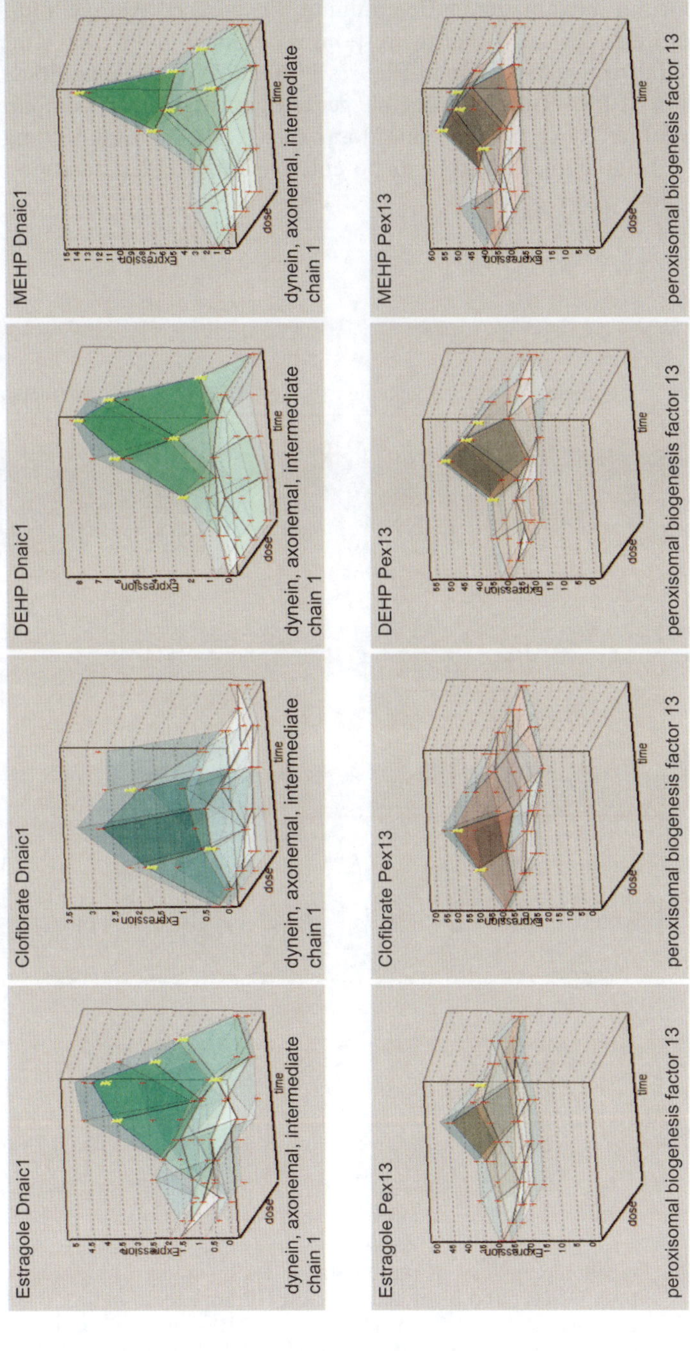

Figure 11.6 The surface of Dnaic1 and Pex13 for estragole, clofibrate, DEHP, and MEHP. The late response may be mediated by Pex13.

peroxisome proliferation may be involved. Indeed, estragole increases liver weight at a dose lower than the carcinogenic dose.[19]

An important advantage in determining the actual average number of mRNA molecules per cell is that the responses obtained in different studies can be compared directly. As shown in Figures 11.6 and 11.7, the magnitude of the up-regulation of PPAR-alpha-inducible genes by estragole was comparable to that of clofibrate, since, at the same doses (*i.e.* 0, 10, 30, and 100 mg kg^{-1}) employed, estragole appears to be as potent as clofibrate in activating PPAR-alpha signaling.

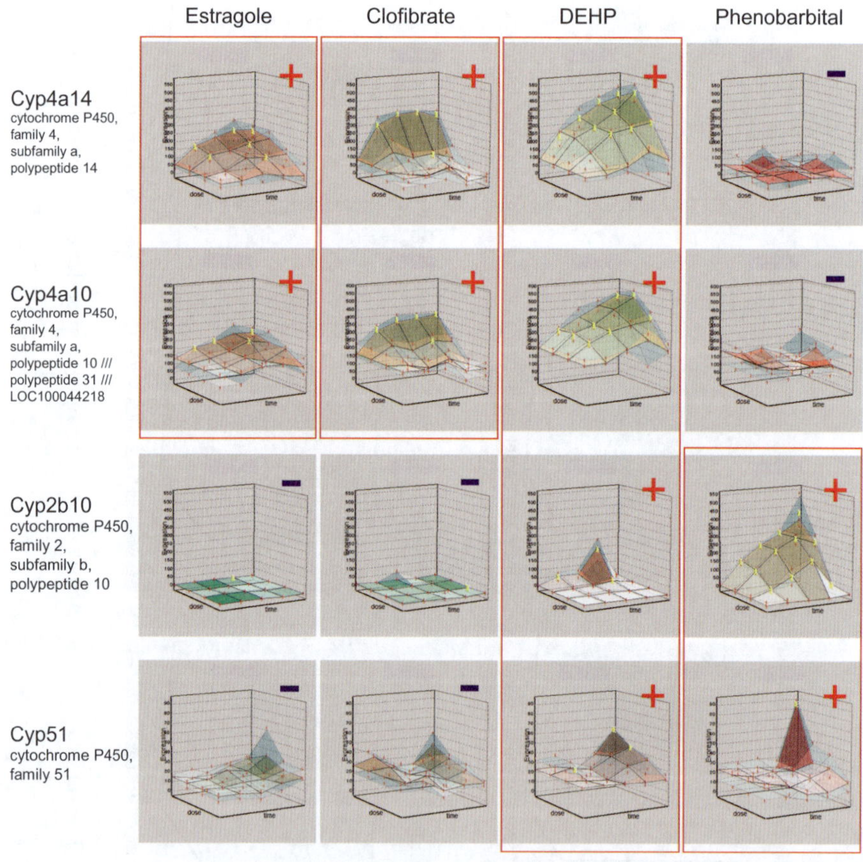

Figure 11.7 Percellome analysis of representative P450s induced by estragole, clofi-brate, DEHP, and phenobarbital. DEHP appears to induce P450s *via* at least two different pathways, *i.e.* PPAR-alpha and CAR, whereas estragole, clofibrate, and phenobarbital induce only PPAR-alpha or CAR, respectively.

11.5 Conclusions

Our present observations that estragole appears to be as potent an agonist of PPAR-alpha as clofibrate (on a mg kg^{-1} basis) should now be confirmed by actual binding and signaling studies. If confirmed, the hepatocarcinogenic potential of this compound should be reevaluated accordingly. Although recent reports on estragole carcinogenicity suggest involvement of its metabolites[20] or glucocorticoid pathways,[21] our Percellome data support neither the involvement of such pathways or pronounced genotoxicity (which can be monitored indirectly as an enhancement in DNA repair and responses to oxidative stress). Interestingly, DEHP and Wyeth 14,643, well-characterized non-genotoxic rodent hepatocarcinogens that evoke tumors through peroxisome proliferation, gave mutation in Lac Z transgenic mice.[22]

Acknowledgements

The authors wish to thank Nae Matsuda, Kenta Yoshiki, Tomoko Ando, Noriko Moriyama, Yuko Kondo, Yuko Nakamura, Maki Abe, Ayako Imai, Koichi Morita, Shinobu Watanabe, Hisako Aihara, and Chiyuri Aoyagi for technical support. This study was supported financially by MHLW Health Sciences Research Grants H18-Kagaku-Ippan-001, H15-Kagaku-002, H14-Toxico-001, and H13-Seikatsu-012.

References

1. H. P. Rusch, B. E. Kline and C. A. Baumann, *Cancer Res.*, 1945, **5**, 431.
2. R. K. Boutwell, M. K. Brush and H. P. Rusch, *Cancer Res.*, 1949, **9**, 741.
3. D. Kritchevsky, M. M. Weber and D. M. Klurfeld, *Cancer Res.*, 1984, **44**, 3174.
4. J. Kanno, K. Aisaki, K. Igarashi, N. Nakatsu, A. Ono, Y. Kodama and T. Nagao, *BMC Genomics*, 2006, **7**, 64.
5. K.-I. Aisaki, 2011, in preparation.
6. J. M. Peters, R. C. Cattley and F. J. Gonzalez, *Carcinogenesis*, 1997, **18**, 2029.
7. S. S. Lee, T. Pineau, J. Drago, E. J. Lee, J. W. Owens, D. L. Kroetz, P. M. Fernandez-Salguero, H. Westphal and F. J. Gonzalez, *Mol. Cell. Biol.*, 1995, **15**, 3012.
8. S. Yu, W. Q. Cao, P. Kashireddy, K. Meyer, Y. Jia, D. E. Hughes, Y. Tan, J. Feng, A. V. Yeldandi, M. S. Rao, R. H. Costa, F. J. Gonzalez and J. K. Reddy, *J. Biol. Chem.*, 2001, **276**, 42485.
9. L. Billiet, C. Furman, C. Cuaz-Perolin, R. Paumelle, M. Raymondjean, T. Simmet and M. Rouis, *J. Mol. Biol.*, 2008, **384**, 564.
10. M. Rakhshandehroo, G. Hooiveld, M. Muller and S. Kersten, *PLoS One*, 2009, **4**, e6796.
11. C. B. Brocard, K. K. Boucher, C. Jedeszko, P. K. Kim and P. A. Walton, *Traffic*, 2005, **6**, 386.

12. C. Kural, H. Kim, S. Syed, G. Goshima, V. I. Gelfand and P. R. Selvin, *Science*, 2005, **308**, 1469.
13. D. T. Odom, N. Zizlsperger, D. B. Gordon, G. W. Bell, N. J. Rinaldi, H. L. Murray, T. L. Volkert, J. Schreiber, P. A. Rolfe, D. K. Gifford, E. Fraenkel, G. I. Bell and R. A. Young, *Science*, 2004, **303**, 1378.
14. H. Ren, L. M. Aleksunes, C. Wood, B. Vallanat, M. H. George, C. D. Klaassen and J. C. Corton, *Toxicol. Sci.*, **113**, 45.
15. J. G. Dekeyser, E. M. Laurenzana, E. C. Peterson, T. Chen and C. J. Omiecinski, *Toxicol. Sci.*, **120**, 381.
16. C. Xu, C. Y. Li and A. N. Kong, *Arch. Pharm. Res.*, 2005, **28**, 249.
17. G. A. Francis, E. Fayard, F. Picard and J. Auwerx, *Annu. Rev. Physiol.*, 2003, **65**, 261.
18. M. Lemoine, J. Capeau and L. Serfaty, *PPAR Res.*, 2009, **2009**, 906167.
19. FAO/WHO, 69th joint meeting, *Safety Evaluation of Certain Food Additives*, WHO Food Additives Series, International Programme on Chemical Safety, World Health Organization, Geneva, 2009.
20. Y. Ishii, Y. Suzuki, D. Hibi, M. Jin, K. Fukuhara, T. Umemura and A. Nishikawa, *Chem. Res. Toxicol.*, **24**, 532.
21. V. I. Kaledin, M. Y. Pakharukova, E. N. Pivovarova, K. Y. Kropachev, N. V. Baginskaya, E. D. Vasilieva, S. I. Ilnitskaya, E. V. Nikitenko, V. F. Kobzev and T. I. Merkulova, *Biochemistry (Moscow)*, 2009, **74**, 377.
22. M. E. Boerrigter, *J. Carcinog.*, 2004, **3**, 7.

CHAPTER 12

Occurrence of Endocrine Disrupters in Food Chains

ALBERTO MANTOVANI[a] AND ILARIA PROIETTI[a,b]

[a] Food and Veterinary Toxicology Unit, Department of Veterinary Public Health and Food Safety, Istituto Superiore di Sanità, Viale Regina Elena 299, 00161 Rome, Italy; [b] Agrisystem, Catholic University of "Sacro Cuore", Via Emilia Parmense 84, 29122 Piacenza, Italy

12.1 Introduction

The general population is exposed to significant levels of a variety of endocrine disrupters (EDs) through the diet, which is therefore a health concern. One major issue is the so-called "cocktail" effect, *i.e.* additive effects of different EDs present at low levels, but interacting with the same targets, *e.g.* nuclear receptors.[1-3] Furthermore, it is not simply the daily dose alone that is of concern, since many EDs bioaccumulate in lipid compartments of tissues, giving rise to a mixed "body burden" of contaminants of different origins, including dioxins, polychlorinated biphenyls, chlorinated pesticides and their metabolites, as well as brominated flame retardants.[4] Some compounds may also be concentrated in food chains (*e.g.* organotins, which can concentrate in seafood), thereby enhancing the overall ED burden.[5]

In general, EDs may be divided into four broad categories from the viewpoint of exposure patterns *via* food chains:

1. EDs that can bioaccumulate in organisms [*e.g.* polychlorinated biphenyls (PCBs), perfluorinated compounds], thereby affecting components of the food chain that are most susceptible to environmental pollution.[6,7]

Issues in Toxicology No. 11
Hormone-Disruptive Chemical Contaminants in Food
Edited by Ingemar Pongratz and Linda Vikström Bergander

Published by the Royal Society of Chemistry, www.rsc.org

2. EDs employed in the production of food, such as agrochemicals[8] and veterinary drugs.[9] In this case, the potential risks for consumers are largely determined by the enactment of up-to-date regulations, as well as enforcement of appropriate farming practices.

3. EDs released into food from contact materials, processing aids, *etc.* The most well-known example of this, recently re-evaluated by the European Food Safety Authority, is bisphenol A, which has caused widespread concern because of its release from polycarbonate-based containers for baby food.[10] Semicarbazide, a less well-known example, is released upon heating from foodstuffs sealed in glass jars, primarily baby foods, and exerts endocrine effects on juvenile rodents, possibly mediated by N-methyl-D-aspartate receptors.[11]

4. The fourth group of EDs or, more accurately, endocrine-active substances, consists of a variety of bioactive compounds naturally present in food. In such cases, the effects may be beneficial or detrimental, depending on the dose, chemical form and endocrine status of the organism (*e.g.* age, gender). For certain essential nutrients, "endocrine disruption" may be caused by a deficiency, as well as by an excess, a paramount example being iodine-deficient hypothyroidism, a major endocrine disorder worldwide.[12]

12.2   Factors Involved in Exposure

Dietary habits are influenced by socioeconomic status, cultural and religious factors and individual choices (*e.g.* vegetarianism/veganism), which may thus exert considerable impact on the intake of both nutrients and contaminants, such as EDs. For instance, extensive consumption of fatty foods of animal origin is associated with greater exposure to persistent EDs.[13,14] It has been estimated that 90% of total human exposure to PCBs and dioxins is through the diet,[6] the contributions of individual items of food depending not only on dietary habits, but also on the environmental quality of the areas of food production. Thus, food preparation, storage and processing must also be taken into account. For instance, management of cereals and nuts before, during and after harvest is critical to the level of contamination by mycotoxins,[15] including zearalenone, a potent estrogenic substance.[16]

12.2.1   Consideration of Different Stages of Life

The health risks related to dietary exposure to EDs varies not only in relationship to different patterns of food consumption, but also at different stages of life.

12.2.1.1   Newborns and Small Infants

Newborns and small infants are exposed to EDs in quite a different manner than adults, since an early body burden may be derived from *in utero* exposure and small infants consume a very limited variety of food, only breast milk in many cases, so that contamination of a single item may determine the safety of their entire diet. Furthermore, internal defences of neonates against contaminants are limited by the immaturity of both hepatic detoxification and the blood/brain barrier.

In the case of breastfeeding, bioaccumulation of a mixture of lipophilic compounds (PCBs, brominated flame retardants, dioxins) and their transfer from the maternal body to the fat fraction of milk is a major concern. Exposure *via* this route can be considerable, *e.g.* it has been estimated that approximately 10% of the body burden of dioxin-like substances at 25 years of age is attributable to breastfeeding during the first six months of life.[17] In assessing the potential risks involved, it is also necessary to consider the advantages of breastfeeding for the child, including improved neurocognitive and behavioural development. For example, breast milk makes an important contribution to neonatal intake of iodine, which is essential to optimal thyroid function and thus to the functional development of the nervous system.[18] Because of these enormous benefits, it is not recommended to reduce exposure to EDs by restricting breastfeeding, except in exceptional circumstances associated with high levels of contamination, *e.g.* by dioxins from e-waste pollution in developing countries.[19]

With regards to formula milk, soy-based products may exert thyrostatic action, probably due to the presence of genistein, a soy isoflavone known to interfere with the synthesis of the thyroid hormone.[20] The exposure of infants fed soy-milk formulas to genistein can be 4- to 5-fold higher than adult exposure. It is recommended, therefore, to use formulas containing only isolated soy proteins, as well as to limit the use of such formulas to specific circumstances (*i.e.* cases of intolerance to proteins in cow's milk, lactose or galactose).

12.2.1.2 Children and Adolescents

Patterns of food consumption vary in a pronounced manner with age, and, for instance, milk may be a more important source of exposure to dioxins and PCBs for children than adults.[13] Moreover, children may be subject to higher exposure simply because they ingest more food in proportion to their body weight, a ratio that slowly decreases during adolescence to become more stable at approximately 20 years of age.[21,22] In addition, childhood is a period of dynamic growth and development, with several windows of specific susceptibility to the effects of EDs on sexual, cognitive and immune development.[23,24] Consequently, a life-stage approach to assessing children's exposure has been proposed.[25]

12.2.1.3 Women of Childbearing Age

With regards to "sustainable food safety", there is general agreement that ED intake by women of childbearing age should be reduced, in order to minimize transgenerational exposure.[26] The numerous critical examples that can be cited include the following:

(a) The simultaneous presence of both hazardous contaminants (methyl-mercury, PCBs) and nutrients essential to embryo–foetal development (iodine, omega-3) in certain types of fish (*e.g.* tuna, cod, salmon) has led to the development of specific measures for risk management and recommendations concerning food choices in certain countries (*e.g.* Canada: http://www.toronto.ca/health/hphe/pdf/boh_fish_mercury.pdf).

(b) Excessive use of "natural" substances with endocrine activity (phytoestrogens, trace elements such as iodine and selenium) due, for example, to incorrect health concerns, should receive more consideration.[27]

12.2.2   Consideration of Specific Dietary Habits

Dietary patterns in Eastern Asia are characterized by high consumption of vegetables, including considerable levels of soybeans, with a high content of phytoestrogens. In Europe, soybean intake varies widely, being greater in northern European countries, which probably reflects the wider spread of vegetarianism in the region.[28] Early and prolonged exposure to phytoestrogens may exert considerable impact on an individual's hormonal homeostasis.[29]

While possibly providing a protective effect against certain diseases, phytoestrogens may prolong the menstrual cycle and, although still controversial, elevate the risk for breast cancer.[30] In addition, in the absence of adequate iodine intake, excessive consumption of soybeans and related products may have an adverse effect on thyroid function.[20] Although such considerations may be of particular concern for vegetarians, a continuously growing group who may have a high intake of isoflavone phytoestrogens,[28] low levels of phytoestrogens, primarily bioactive metabolites such as enterolignans and equol, may be found in all foods of animal origin.[31] At the same time, vegetarians are probably less exposed to other lipophilic EDs, since a recent study on a large Italian population[13] revealed that fish and fish products and milk and dairy products are the major dietary sources of dioxin-like and non-dioxin-like PCBs. Clearly, vegetarians deserve specific attention in this context.

A study conducted in the USA in 1997 revealed that the intake of dioxin-like substances and DDE through fast-food meals accounts for 16.7–52.7% of the daily TEQ (international toxic equivalent).[32] If confirmed, this finding indicates that, in relationship to body weight, consumption of dioxins *via* fast food is almost three times higher in children and adolescents than in adults.

12.2.3   Consideration of Certain Specific Items of Food

12.2.3.1   Cereals

Cereals often receive relatively little attention as a possible source of contaminants, which is unfortunate since cereals constitute a major food staple worldwide. For assessment and management of ED exposure through cereals and their products, the following geomedical and agricultural aspects, among others, are of major relevance: (a) mycotoxin contamination, in particular by zearalenone, a powerful estrogenic ED which can be present in all types of cereals, especially corn;[16] (b) specific accumulation, for example enrichment of arsenic, a potential diabetogenic ED, in rice;[33] and (c) the contribution of toxic or protective trace elements by different types of soil (*e.g.* the significant differences in the selenium content of agricultural products even from neighbouring geographic areas in Italy).[34,35]

12.2.3.2   Fruits and Vegetables

Although widely recognized as a source of beneficial vitamins, trace elements, fibre and antioxidants, fruits and vegetables are also considered to be a major source of pesticide residues. Actually, monitoring programmes indicate only low levels of pesticide residues in vegetables produced in Europe, although the corresponding levels in products from non-EU countries are higher.[36] Some contemporary investigations do indicate the presence of several pesticide metabolites (including metabolites of organophosphates and ethylene bisdithiocarbamates) in children and adults with no known occupational exposure,[37,38] emphasizing the importance of assessing exposure through all routes (diet, living environment, consumer products) as well as cumulative effects. Indeed, several groups of pesticides have common metabolites that can serve as indicators of cumulative exposure. For example, urinary ethylene thiourea and the thyrostatic metabolite of ethylene bisdithiocarbamates are biomarkers of exposure to this group of compounds.[8] Information concerning the impact of combined exposure to different substances on human health is scarce and warrants greater attention. For instance, cumulative prenatal exposure to organophosphates is associated with problems in neurocognitive developmental during early childhood.[39]

In addition to phytoestrogens, many other endocrine-active compounds may be present in vegetables. Most known vegetable goitrogens are cyanogenic compounds, such as thiocyanates and isothiocyanates derived from glucosinolates ingested in, among other items, Brassicaceae (*Brassica* spp.). The presence of cyanogenic compounds in animal feed is well known, whereas their impact on human health is probably small, at least at current levels of consumption in Europe.[40]

Many flavonoids present in fruits and vegetables are powerful inhibitors of sulfotransferase enzymes involved in the removal and detoxification of xenobiotics and essential to the metabolism of steroid and thyroid hormones.[41] In this context, furocoumarins, present in grapefruit and other citrus fruits, are capable of mediating dioxin-like effects by interaction with the aryl hydrocarbon receptor (AhR) and are of growing concern. However, the risk to healthy consumers appears at present to be low (probably due to the action of intestinal bacterial flora and other detoxifying pathways), although there is a realistic possibility that the effects of certain drugs might be altered by inhibitors of sulfotransferases.[42]

With regards to environmental EDs, a recent study in Italy[13] confirmed that vegetables and fruits make only a very small contribution (3%) to the total dietary intake of dioxin-like and non-dioxin-like PCBs. In contrast, vegetable oils account for 11–16% of the total dietary intake of PCBs, a finding of obvious relevance to all with a Mediterranean diet.

At the same time, fruits and vegetables provide dietary fibre, which is important in modulating the activity of intestinal microflora, as well as the absorption of dietary components. Indeed, the bioavailability and intestinal metabolism of estrogens is influenced by the fat/fibre ratio in the diet.[43]

12.2.3.3   Milk and Dairy Products

Milk and dairy products contribute to dietary exposure to EDs to a degree that differs considerably at different ages, being particularly high during childhood. Contaminants in feed and fodder are rapidly absorbed and accumulate in the liver, adipose tissue and milk fat of milk-producing animals. A 2005 longitudinal French evaluation of the levels of PCDDs, dibenzofurans and PCBs in milk and milk products[44] revealed a reduction of 50% compared to values for 1998, to levels now below the maximal limits recommended by the EU. In Italy, milk accounts for 38% and 24% of the total dietary intake of dioxins and dioxin-like compounds by children and adults, respectively.[13] In certain regions, illegal handling of industrial wastes may lead to prolonged contamination of pastures, with consequent severe risk to livestock and enhanced consumer exposure to persistent EDs, such as PCBs.[45] Contamination of pastures exposed to sewage by phthalates is also possible.[46] In the case of PCBs and related compounds, the lipid content of ingested milk and dairy products also matters. Habitual consumption of partially skimmed milk in infancy may reduce exposure to dioxin-like compounds and the resulting body burden by 10–20%.[47]

12.2.3.4   Seafood

In addition to the well-known contamination by PCBs and dioxins, certain types of seafood may contain other contaminants, including brominated flame retardants (BFRs) and perfluorinated compounds that do sum up with the more known persistent pollutants.[7,48] Levels of organotins are especially high in molluscs, but all seafood is a potential source.[5] Benthic seafood organisms also contain especially high levels of nonylphenols[49] and cadmium.[50] With respect to risk assessment concerning heavy metals, species differences must be taken into consideration. For instance, in fish, arsenic is largely present as arsenobetaine (of minimal toxicity) and other little-studied organoarsenical compounds, arsenolipids or arsenosugars,[51] so that in this case the determination of total arsenic content does not provide an accurate assessment of the actual risk to the consumer.

12.2.3.5   The Feed–Food Chain

Components of animal feed are a prime focus of safety assessment in the "farm-to-fork" approach.[52] Feedstuffs can be a major sources of persistent EDs, such as PCBs, *e.g.* in connection with fish farming and grazing in polluted areas.[45] In addition, certain feed additives (*e.g.* organic forms of chromium, not yet authorized in Europe) may modulate endocrine homeostasis, with possible deleterious effects on farm animals and consumers. However, the availability of information concerning the feed-to-food transfer of major emerging groups of EDs, such as BFRs and perfluorinated compounds, remains highly limited.[53] Although EDs of vegetable origin (*e.g.* zearalenone and isothiocyanates) may exert more pronounced adverse impacts on farm animals than do environmental xenobiotics, the risk of transfer to consumers appears to be low.[16,40] The issue of EDs in the feed–food chain still presents many uncertainties and more refined exposure data, characterization of biomarkers in farm animals and a search for feed less liable to contamination are necessary.[53]

12.3   Interactions between Endocrine Disrupters and Components of Food

An up-to-date approach to food toxicology cannot simply consider the diet as a source of exposure to external harmful substances. Contaminants such as EDs may interact with the same metabolic pathways as natural food components, including polyunsaturated fatty acids, trace elements and vitamins, as well as other bioactive substances (*e.g.* polyphenols) that cannot be considered nutrients since deficiency is not associated with any known deleterious effects.[54] Certain items of food may be rich in nutrients, as well as vulnerable to accumulation of contaminants, thus motivating a food-specific balanced evaluation of toxicological risks and nutritional benefits.

One relevant example is represented by the salmonids and other seafood that provide useful nutrients (*e.g.* polyunsaturated fatty acids, selenium and iodine), but are, at the same time, a major source of EDs and other bioaccumulating contaminants, such as methylmercury.[14] Thus, the available evidence might be interpreted to justify recommendations to increase or reduce fish consumption, which puts risk managers in a very awkward situation. Reduction of fish consumption may not be necessary in Europe today, but monitoring of contaminants in edible fish should be continued, as should the development of novel, less-contaminated aquaculture feeds.[55]

Most interestingly, contaminants and natural components of food may interact with the same pathways and targets to produce complex outcomes, depending on the dose and particular target. For example, phytoestrogens protect against certain hormone-dependent cancers, as well as postmenopausal osteoporosis, but may also interfere with receptor-mediated signal transduction (*e.g.* by inhibiting protein kinase) and DNA replication.[56]

12.3.1   Interactions between Xenobiotics and "Natural" Substances

To date, scientific investigations on interactions between xenobiotics and "natural" substances in food are still limited in number, but several relevant examples are provided below.

12.3.1.1 Iodine and Endocrine Disruptors

The main dietary sources of iodine, the major determinant of thyroid development and function, are seafood and milk. Subclinical iodine deficiency is still a common problem in many areas, including Europe.[27] The thyroid gland is also increasingly recognized as a primary target for EDs, including novel ones such as organophosphorus insecticides.[57] Nonetheless, only a few publications have focused on low iodine status in relationship to susceptibility to xenobiotics.

Somewhat unexpectedly, the widespread phthalate plasticizers, best known as anti-androgens, can modulate basal iodide uptake *via* the sodium/iodide symporter in thyroid follicular cells *in vitro*. This effect is not shared by all phthalates and is independent of cytotoxicity.[58] Furthermore, many phytoestrogens may interfere with iodination of thyroid hormones, some (*e.g.* naringenin and quercetin, which contain a resorcinol moiety) by direct and potent inhibition of thyroid peroxidise and others by inhibition (myricetin, naringin) or competitive inhibition (biochanin A) of tyrosine iodination.[59] A Czech biomonitoring study also indicated an adverse effect of genistein on thyroid function in children, an effect of modest severity, but, nonetheless, potentially significant when iodine intake is insufficient.[20] Perchlorate, a contaminant of drinking water, inhibits thyroidal uptake of iodide, an effect that was more pronounced in iodine-deficient than in well-nourished female rats.[60]

Thus, experimental investigations suggest that interactions between iodine and certain thyroid-specific EDs may be less straightforward than predicted. On the other hand, limited studies on humans hint at a protective effect. Infants exposed to high levels of PCBs and hexachlorobenzene through fish in their diet demonstrated only modest changes in thyroid parameters, possibly due to a protective effect of the iodine also present in fish.[61] Perchlorate, employed as a solvent and fertilizer and formed as a by-product of chlorination of water, was found in a Chilean survey not to affect markers of thyroid function in iodine-sufficient women subjected to long-term exposure through drinking water during pregnancy and lactation.[62]

12.3.1.2 Phytoestrogens and "Xeno" Endocrine Disruptors

Phytoestrogens can be viewed as a sort of "naturally occurring EDs" and the most extensively investigated of these is genistein, which is present in soy and, thus, at high levels in Eastern Asian diets. In addition, overall intake of phytoestrogens through European diets may also be significant.[28,63] Flavonoids (daidzein, genistein, quercetin and luteolin) can partially antagonize the stimulation of proliferation of estrogen-dependent MCF-7 human breast cancer cells caused by synthetic estrogenic EDs, including alkylphenols (by-products of anionic detergent production), the plastic additive bisphenol A, and the PCB 4,4'-dihydroxybiphenyl.[64,65] Such findings indicate that phytoestrogens can compete with other estrogenic EDs for common biological targets, thus exerting protective action. However, in other model systems, no such interaction was observed. For example, genistein did not modulate the effects of two persistent EDs, the polybrominated flame retardant PBDE-99 and the PCB mixture Aroclor 1254, on human astroglial cells.[66]

As is sometimes the case, the *in vivo* findings are more complex. Genistein and the estrogenic chlorinated insecticide methoxychlor exerted an additive impact on both immune functions and immunological development in rats, indicating that the developing thymus is a sensitive target for such combined exposure.[67] In estrogen reporter (ERE-tK-luciferase) male mice, genistein modulated the actions of both estradiol and persistent EDs in a tissue-specific manner: the anti-estrogenic actions of β-hexachlorocyclohexane on the testis and of *o,p'*-DDT on the liver were antagonized, whereas the effect of genistein on the liver was additive with that of the ER agonist *p,p'*-DDT on this same organ.[68]

When two defined mixtures of phytoestrogens and "xeno" EDs (designed on the basis of human exposure data) were tested in the uterotrophic assay on prepubertal rats, the phytoestrogen mixture alone elicited an uterotrophic response, whereas the synthetic mixture exerted an effect only in combination with the phytoestrogens, possibly because the levels of exposure were too low.[69] Accordingly, combined exposure to estrogenic and anti-androgenic EDs may represent a potential risk to male reproductive development. The indices of hypospadias in mice exposed to genistein, vinclozolin or both were 25%, 42% and 41%, respectively, indicating an effect that was less than additive.[70] On the other hand, genistein, as well as the methyl donor folic acid, both antagonize the hypomethylation effect of DNA caused by bisphenol A in mouse embryos.[71] Thus, available data indicate that interactions between phytoestrogens and EDs may be important but complex, showing additivity or antagonism, depending on the specific compounds involved, end-points determined and age of the organism.

12.3.1.3 *Endocrine Disruptors and Vitamin A*

All-*trans*- and 9-*cis*-retinoic acid are metabolites of vitamin A that interact with the nuclear receptors RAR and RXR, respectively. The pathways in which retinoic acid participates cross-talk with those involving the aryl hydrocarbon receptor (AhR), the direct cellular target for dioxins and dioxin-like compounds.[72] Dioxins are potent up-regulators of cytochrome P450 (CYP) 1A1, a phenomenon that may, in turn, enhance the effects of dioxin, whereas concurrent supplementation with vitamin A lowers the levels of CYP1A1, mRNA, protein and activity in dioxin-exposed mice, thereby attenuating liver damage.[73] Mice lacking retinoid-binding proteins, in particular retinoid-binding protein I, are especially susceptible to dioxin-induced depletion of hepatic retinoids.

In general, RAR- and RXR-knockout mice are essentially as sensitive to such depletion as wild-type animals, with the exception of RXRbeta–/– mice, which exhibit no decrease in hepatic levels of vitamin A upon exposure to dioxin, suggesting a possible role of this receptor in this phenomenon.[74]

In addition, retinoid storage and metabolism are disrupted in two strains of female rats with differing sensitivities to dioxin (Long-Evans and Han/ Wistar).[75] Comparison of the effect of dioxin on hepatic levels of retinyl palmitate in AhR+/– and AhR–/– mice confirms that disruption of retinoid homeostasis is a primary AhR-mediated mode of action by dioxin-like chemicals.[76]

Moreover, pathways involving retinoids may be critical targets for polybrominated diphenyl ethers as well. In rats treated orally with penta-BDE-71, a reduced levels of hepatic apolar retinoids, together with lowered serum levels of thyroid hormone, were the most pronounced effects observed.[77] Such studies also suggest that vitamin A deficiency might aggravate susceptibility to certain persistent EDs.

Although the scientific evidence is still quite limited, several other examples are documented in the Endocrine-disrupting chemicals–Diet Interaction Database (EDID; the only database dedicated at present to ED-nutrient interactions).[54] One such is the general protection exerted by the "antioxidant" vitamins C and E against the effects of several EDs, including dioxin-like polychlorinated biphenyls and phthalates. Indeed, the modes of action of several EDs appear to eventually enhance oxidative stress.[78] Overall, new information concerning interactions between EDs and natural components of food may reveal novel food-related factors that modulate vulnerability, as well as nutrient intake, for use in risk prevention and/or risk reduction strategies.

12.3.1.4 *Anti-nutritional Factors and Endocrine Homeostasis*

Anti-nutritional factors (ANF), such as phytic acid (inositol hexaphosphate) and tannins, can chelate metal cations, such as iron, zinc and copper, as well as proteins to form insoluble precipitates, thereby reducing the bioavailability of trace minerals and the digestibility of proteins.[79–81]

Recently, much attention has been focused on the potential of heavy metals to disrupt signalling pathways in humans. In particular, several metallic compounds (*e.g.* organotins and compounds containing cadmium, mercury, arsenic, lead or manganese) exhibit endocrine activities, such as estrogenicity, both *in vitro* and *in vivo*.[82–84] In addition, the essential trace elements zinc and copper potentiate estradiol-induced responses in a dose-dependent manner, at least *in vitro*.[85]

Moreover, zinc deficiency is associated with poor semen quality. For instance, an epidemiological study in China found that serum levels of zinc ≥ 1 mg L^{-1} were associated with enhanced male fertility, exerting a positive influence on spermatogenesis and reducing the risk of asthenozoospermia.[86] This protective role probably reflects the ability of zinc to maintain sperm viability by inhibiting DNAases, as well as its membrane-stabilizing and anti-oxidant activities.[84,87] Furthermore, in seminal plasma, zinc appears to protect against superoxide anions, both by acting as a scavenger of this highly reactive oxygen species produced by defective spermatozoa in the semen after ejaculation and by inhibiting its production by superoxide dismutase-like activity in spermatozoa in a dose-dependent manner *in vitro*.[88]

Copper and zinc deficiency also disrupt female reproductive functions. Copper-zinc superoxide dismutase acts as a scavenger of superoxide radicals and stimulates progesterone production in the corpus luteum.[89,90] In addition, the balance between zinc and copper influences thyroid homeostasis: zinc is involved in the conversion of the inactive T4 (thyroxine hormone) into active T3 (triiodothyronine hormone), whereas copper stimulates the production of T4 and controls body levels of calcium, which prevents excessive uptake of T4 into blood cells. In the case of copper deficiency, excessive hormone is produced by the thyroid, resulting in hyperthyroidism, while excessive copper intake leads to hypothyroidism.[91,92]

In protein-deficient rats, zinc protects against oxidative stress and modulates hepatic concentrations of endocrine-active trace elements, such as copper and selenium.[93] Moreover, zinc plays a protective role with respect to several EDs. A number of *in vivo* investigations have demonstrated the ability of dietary zinc to alleviate the toxic effects of arsenic, nickel and chlorpyrifos on rat liver,[94–96] as well as to protect against cadmium-induced testicular toxicity.[97–99] Cadmium inhibits androgen production in male rats, by damaging seminiferous tubules and causing degeneration and disintegration of spermatogenic cells and both pre- and co-treatment with zinc protect by restoring key steroidogenic enzymes and testicular protein levels, as well as increasing testicular weight.[97,100]

The level of iron can also influence the susceptibility of rats to certain EDs. Recently, Rashid and colleagues[101] reported that iron deficiency augments lipid peroxidation, attenuates glutathione levels and modulates tissue antioxidant enzymes in rats exposed to the estrogenic ED bisphenol A. Accordingly, iron deficiency may aggravate the oxidative stress induced by several EDs. Although the reasons remain unclear, this modulatory effect of iron deficiency is more pronounced in female than in male rats.

By binding to divalent cations of zinc, copper and iron, the anti-nutritional factors phytic acid and tannins may exert indirect adverse effects on endocrine homeostasis.[102–104] Indeed, pronounced consumption of items of food containing high levels of phytic acid and/or tannins can result in mineral deficiencies. Phytic acid is a major component of grains, legumes, roots and tubers, where it serves as a source of phosphate for germination and growth.[105,106] Tannins are naturally occurring plant polyphenols which serve as defences against pathogens, insects and herbivores.[107–109] Both are widely distributed in the plant kingdom, although at variable levels, being present in many types of plants and plant tissues (*e.g.* tree bark, wood, leaves, roots, fruits, legumes and grasses).[107]

Because of this broad distribution, the potential effects of phytic acid and tannins on endocrine homeostasis represent a major health concern, especially for populations whose diet relies heavily on cereals and legumes and is not very varied, as in developing countries.[102,104]

The cereal sorghum is one example of a staple food in developing countries characterized by potentially good nutritional value, but also containing high levels of ANF.[110,111] While sorghum contains ample amounts of micronutrients (its mean contents of iron and zinc being 5.8 and 2.1 mg/100 g, respectively),[110] its high content of phytic acid and tannins impairs the bioavailability of these trace elements in the human organism.[102–104,111] In developing countries this results in secondary micronutrient deficiency, which influences endocrine homeostasis[102,104] as well as potentially enhancing susceptibility to environmental EDs.[94–101]

12.4 Conclusions

Although the diet is recognized as a major route of exposure to EDs, risk assessment in this context currently involves a number of uncertainties. Appropriate environmental monitoring, the use of "cleaner" foods (feed quality significantly influences the levels of contamination in, *e.g.*, fish, milk and dairy products) and effective exchange of information with food producers are essential to reducing risks while at the same time maintaining adequate dietary availability, variety and nutrition. In connection with modern food safety, both responsible and informed choices by customers and technological innovation are crucial. For instance, current systems for control of food hygiene, which focus mainly on microbiological risks (HACCP: Hazard Analysis Critical Control Point), could be extended to the prevention of long-term toxicological risks associated with food chains.[26]

Controls are necessary, but cannot be sufficient, even if official monitoring programmes are continuously updated to include additional emerging contaminants. Prevention should focus on primary production. The establishment of maximal acceptable residues (*e.g.* of pesticides and feed additives) or tolerable levels (of contaminants) aims mainly at making production safety, but cannot, at present, guarantee protection against prolonged exposure to low levels of multiple, potentially additive EDs. Cost-effective strategies for reducing exposure to EDs should be directed towards the sources of contamination of food and animal feed. Public bodies should both enforce the application and necessary updating of current regulations and also promote technological innovation designed to achieve safer production, as well as an integrated management of risks. Indeed, the "Hygiene Package" (EC 852/2004, 853/2004, 854/2004 and 882/04; http://www.europass.parma.it/page.asp?IDCategoria=1117&IDSezione=7855) assigns primary responsibility for the maintenance of a high level of food chain safety to the producers. At the same time, much more scientific effort should be focused on risk–benefit assessment of food commodities as a whole.

Acknowledgments

This article was written within the framework of the "PREVIENI" project, supported by the Italian Ministry for Environment (http://www.iss.it/prvn), as well as the PhD project "Effects of fermentation on phytic acid and tannins activities in the sorghum porridge, assessed by the *in vitro* model DLD-1, and study of its shelf-life through biosensor techniques" (Ilaria Proietti, AgriSystem, Catholic University of Piacenza, Italy).

References

1. P. R. Jacobsen, S. Christiansen, J. Boberg, C. Nellemann and U. Hass, *Int. J. Androl.*, 2010, **33**, 434–442.
2. M. B. Kjaerstad, C. Taxvig, H. R. Andersen and C. Nellemann, *Int. J. Androl.*, 2010, **33**, 425–433.
3. J. L. Flippin, J. M. Hedge, M. J. De Vito, G. A. Leblanc and K. M. Crofton. *Int. J. Toxicol.*, 2009, **28**, 368–381; erratum, *Int. J. Toxicol.*, 2010, **29**, 135.
4. H. Fromme, M. Albrecht, J. Angerer, H. Drexler, L. Gruber, M. Schlummer, H. Parlar, W. Korner, A. Wanner, D. Heitmann, E. Roscher and G. Bolte, *Int. J. Hyg. Environ. Health*, 2007, **210**, 345–349.
5. European Food Safety Authority, *EFSA J.*, 2004, **102**, 1–119.
6. European Food Safety Authority, *EFSA J.*, 2005, **284**, 1–137.
7. European Food Safety Authority, *EFSA J.*, 2008, **653**, 1–131.
8. A. Mantovani, F. Maranghi, C. La Rocca, G. M. Tiboni and M. Clementi, *Reprod. Toxicol.*, 2008, **26**, 1–7.
9. A. Mantovani and A. Macrì, *J. Exp. Clin. Cancer Res.*, 2002, **21**, 445–456.
10. European Food Safety Authority, *EFSA J.*, 2010, **8**, 1829.
11. F. Maranghi, R. Tassinari, D. Marcoccia, I. Altieri, T. Catone, G. De Angelis, E. Testai, S. Mastrangelo, M. G. Evandri, P. Bolle and S. Lorenzetti, *Chem. Biol. Interact.*, 2010, **183**, 40–48.
12. M. Andersson, B. de Benoist and L. Rogers, *Best Pract. Res. Clin. Endocrinol. Metab.*, 2010, **24**, 1–11.
13. E. Fattore, R. Fanelli, E. Dellatte, A. Turrini and A. di Domenico, *Chemosphere*, 2008, **73**, S278–S283.
14. D. Mozaffarian and E. B. Rimm, *J. Am. Med. Assoc.*, 2006, **296**, 1885–1899.
15. B. Kabak, A. D. Dobson and I. Var, *Crit. Rev. Food Sci. Nutr.*, 2006, **46**, 593–619.
16. European Food Safety Authority, *EFSA J.*, 2004, **89**, 1–35.
17. S. Patandin, P. C. Dagnelie, P. G. Mulder, E. Op de Coul, J. E. van der Veen, N. Weisglas-Kuperus and P. J. Sauer, *Environ. Health Perspect.*, 1999, **107**, 45–51.
18. F. Azizi and P. Smyth, *Clin. Endocrinol.*, 2009, **70**, 803–809.
19. C. Frazzoli, O. E. Orisakwe, R. Dragone and A. Mantovani, *Environ. Impact Assess. Rev.*, 2010, **30**, 388–399.
20. J. Milerová, J. Cerovská, V. Zamrazil, R. Bílek, O. Lapcík and R. Hampl, *Clin. Chem. Lab. Med.*, 2006, **44**, 171–174.
21. E. A. Cohen Hubal, L. S. Sheldon, J. M. Burke, T. R. McCurdy, M. R. Berry, M. L. Rigas, V. G. Zartarian and N. C. Freeman, *Environ. Health Perspect.*, 2000, **108**, 475–486.
22. WHO, *Environmental Health Criteria 237*, World Health Organization, Geneva, 2007.
23. L. M. Schell and M. V. Gallo, *Physiol. Behav.*, 2010, **99**, 246–253.

24. G. Latini, G. Knipp, A. Mantovani, M. Loredana, F. Chiarelli and O. Söder, *Mini Rev. Med. Chem.*, 2010, **10**, 846–855.
25. E. A. Cohen Hubal, J. Moya and S. G. Selevan, *Birth Defects Res. B, Dev. Reprod. Toxicol.*, 2008, **83**, 522–529.
26. C. Frazzoli, C. Petrini and A. Mantovani, *Ann. Ist. Super. Sanità*, 2009, **45**, 65–75.
27. M. G. Soni, T. S. Thurmond, E. R. Miller 3rd, T. Spriggs, A. Bendich and S. T. Omaye, *Toxicol. Sci.*, 2010, **118**(2), 348–355.
28. P. H. Peeters, N. Slimani, Y. T. van der Schouw, P. B. Grace, C. Navarro, A. Tjonneland, A. Olsen, F. Clavel-Chapelon, M. Touillaud, M. C. Boutron-Ruault, M. Jenab, R. Kaaks, J. Linseisen, A. Trichopoulou, D. Trichopoulos, V. Dilis, H. Boeing, C. Weikert, K. Overvad, V. Pala, D. Palli, S. Panico, R. Tumino, P. Vineis, H. B. Bueno-de-Mesquita, C. H. van Gils, G. Skeie, P. Jakszyn, G. Hallmans, G. Berglund, T. J. Key, R. Travis, E. Riboli and S. A. Bingham, *J. Nutr.*, 2007, **137**, 294–300.
29. E. R. Ball, M. K. Caniglia, J. L. Wilcox, K. A. Overton, M. J. Burr, B. D. Wolfe, B. J. Sanders, A. B. Wisniewski and C. C. Wrenn, *Horm. Behav.*, 2010, **57**, 313–322.
30. H. B. Patisaul and W. Jefferson, *Front. Neuroendocrinol.*, 2010, **31**, 400–419.
31. G. G. Kuhnle, C. Dell'Aquila, S. M. Aspinall, S. A. Runswick, A. A. Mulliganand and S. A. Bingham, *J. Agric. Food Chem.*, 2008, **56**, 10099–10104.
32. A. Schecter and L. Li, *Chemosphere*, 1997, **34**, 1449–1457.
33. J. R. Peralta-Videa, M. L. Lopez, M. Narayan, G. Saupe and J. Gardea-Torresdey, *Int. J. Biochem. Cell Biol.*, 2009, **41**, 1665–1677.
34. M. Spadoni, M. Voltaggio, M. Carcea, E. Coni, A. Raggi and F. Cubadda, *Sci. Total Environ.*, 2007, **376**, 160–177.
35. F. Cubadda, S. Ciardullo, M. D'Amato, A. Raggi, F. Aureli and M. Carcea, *J. Agric. Food Chem.*, 2010, **58**, 10176–10183.
36. Commission of the European Communities, 2006; available from: eu/food/fvo/specialreports/pesticide residues/report 2006_en.pdf.
37. C. Aprea, M. Strambi, M. T. Novelli, L. Lunghini and N. Bozzi, *Environ. Health Perspect.*, 2000, **108**, 521–525.
38. C. Saieva, C. Aprea, R. Tumino, G. Masala, S. Salvini, G. Frasca, M. C. Giurdanella, I. Zanna, A. Decarli, G. Sciarra and D. Palli, *Sci. Total Environ.*, 2004, **332**, 71–80.
39. B. Eskenazi, A. R. Marks, A. Bradman, K. Harley, D. B. Barr, C. Johnson, N. Morga and N. P. Jewell, *Environ. Health Perspect.*, 2007, **115**, 792–798.
40. European Food Safety Authority, *EFSA J.*, 2008, **590**, 1–76.
41. R. M. Harris and R. H. Waring, *Curr. Drug Metab.*, 2008, **9**, 269–275.
42. K. van Ede, A. Li, E. Antunes-Fernandes, P. Mulder, A. Peijnenburg and R. Hoogenboom, *Anal. Chim. Acta*, 2008, **617**, 238–245.
43. M. Aubertin-Leheudre, S. Gorbach, M. Woods, J. T. Dwyer, B. Goldin and H. Adlercreutz, *J. Steroid Biochem. Mol. Biol.*, 2008, **112**, 32–39.

44. B. Durand, B. Dufour, D. Fraisse, S. Defour, K. Duhem and K. Le Barillec, *Chemosphere*, 2008, **70**, 689–693.
45. C. La Rocca and A. Mantovani, *Ann. Ist. Super. Sanita*, 2006, **42**, 410–416.
46. S. M. Rhind, C. E. Kyle, C. Mackie and G. Telfer, *Sci. Total Environ.*, 2007, **383**, 70–80.
47. A. L. Yaktine, G. G. Harrison and R. S. Lawrence, *Nutr. Rev.*, 2006, **64**, 403–409.
48. N. Borghesi, S. Corsolini, P. Leonards, S. Brandsma, J. de Boer and S. Focardi, *Chemosphere*, 2009, **77**, 693–698.
49. F. Ferrara, N. Ademollo, M. Delise, F. Fabietti and E. Funari, *Chemosphere*, 2008, **72**, 1279–1285.
50. G. Falco, J. M. Llobet, A. Bocio and J. L. Domingo, *J. Agric. Food Chem.*, 2006, **54**, 6106–6112.
51. S. Ciardullo, F. Aureli, E. Coni, E. Guandalini, F. Iosi, A. Raggi, G. Rufo and F. Cubadda, *J. Agric. Food Chem.*, 2008, **56**, 2442–2451.
52. A. Mantovani, F. Maranghi, I. Purificato and A. Macrì, *Ann. Ist. Super. Sanità*, 2006, **42**, 427–432.
53. A. Mantovani, C. Frazzoli and C. La Rocca, *Vet. J.*, 2009, **182**, 392–401.
54. F. Baldi and A. Mantovani, *Ann. Ist. Super. Sanità*, 2008, **44**, 57–63.
55. European Food Safety Authority, *EFSA J.*, 2005, **236**, 1–118.
56. J. H. Martin, S. Crotty and P. N. Nelson, *Future Oncol.*, 2007, **3**, 307–318.
57. S. De Angelis, R. Tassinari, F. Maranghi, A. Eusepi, A. Di Virgilio, F. Chiarotti, L. Ricceri, A. Venerosi Pesciolini, E. Gilardi, G. Moracci, G. Calamandrei, A. Olivieri and A. Mantovani, *Toxicol. Sci.*, 2009, **108**, 311–319.
58. A. Wenzel, C. Franz, E. Breous and U. Loos, *Mol. Cell. Endocrinol.*, 2005, **244**, 63–71.
59. R. L. Divi and D. R. Doerge, *Chem. Res. Toxicol.*, 1996, **9**, 16–23.
60. B. F. Paulus, M. A. Bazar, C. J. Salice, D. R. Mattie and M. A. Major, *J. Toxicol. Environ. Health A*, 2007, **70**, 1142–1149.
61. R. Dallaire, E. Dewailly, P. Ayotte, G. Muckle, C. Laliberte and S. Bruneau, *Environ. Res.*, 2008, **108**, 387–392.
62. R. Téllez, P. Michaud Chacón, C. Reyes Abarca, B. C. Blount, C. B. Van Landingham, K. S. Crump and J. P. Gibbs, *Thyroid*, 2005, **15**, 963–975.
63. N. M. Saarinen, C. Bingham, S. Lorenzetti, A. A. Mortensen, S. Mäkela, P. Penttinen, I. K. Sorensen, L. M. Valsta, F. Virgili, G. Vollmer, A. Wärri and O. Zierau, *Genes Nutr.*, 2006, **1**, 143–158.
64. D. H. Han, M. S. Denison, H. Tachibana and K. Yamada, *Biosci. Biotechnol. Biochem.*, 2002, **66**, 1479–1487.
65. N. Rajapakse, E. Silva, M. Scholze and A. Kortenkamp, *Environ. Sci. Technol.*, 2004, **38**, 6343–6352.
66. F. Madia, G. Giordano, V. Fattori, A. Vitalone, I. Branchi, F. Capone and L. G. Costa, *Toxicol. Lett.*, 2004, **154**, 11–21.

67. T. L. Guo, X. L. Zhang, E. Bartolucci, J. A. McCay, K. L. White Jr. and L. You, *Toxicology*, 2002, **172**, 205–215.
68. M. Penza, C. Montani, A. Romani, P. Vignolini, P. Ciana, A. Maggi, B. Pampaloni, L. Caimi and D. Di Lorenzo, *Toxicol. Sci.*, 2007, **97**, 299–307.
69. J. A. van Meeuwen, M. van den Berg, J. T. Sanderson, A. Verhoef and A. H. Piersma, *Toxicol. Lett.*, 2007, **170**, 165–176.
70. M. L. Vilela, E. Willingham, J. Buckley, B. C. Liu, K. Agras, Y. Shiroyanagi and L. S. Baskin, *Urology*, 2007, **70**, 618–621.
71. D. C. Dolinoy, D. Huang and R. L. Jirtle, *Proc. Natl. Acad. Sci. USA*, 2007, **104**, 13056–13061.
72. K. A. Murphy, L. Quadro and L. A. White, *Vitam. Horm.*, 2007, **75**, 33–67.
73. Y. M. Yang, D. Y. Huang, G. F. Liu, J. C. Zhong, K. Du, Y. F. Li and X. H. Song, *Toxicol. Sci.*, 2005, **85**, 727–734.
74. P. Hoegberg, C. K. Schmidt, N. Fletcher, C. B. Nilsson, C. Trossvik, A. Gerlienke Schuur, A. Brouwer, H. Nau, N. B. Ghyselinck, P. Chambon and H. Hakansson, *Chem. Biol. Interact.*, 2005, **156**, 25–39.
75. N. Fletcher, N. Giese, C. Schmidt, N. Stern, P. M. Lind, M. Viluksela, J. T. Tuomisto, J. Tuomisto, H. Nau and H. Hakansson, *Toxicol. Sci.*, 2005, **86**, 264–272.
76. N. Nishimura, J. Yonemoto, Y. Miyabara, Y. Fujii-Kuriyama and C. Tohyama, *Arch. Toxicol.*, 2005, **79**, 260–267.
77. L. T. van der Ven, T. van de Kuil, A. Verhoef, P. E. Leonards, W. Slob, R. F. Cantón, S. Germer, T. Hamers, T. J. Visser, S. Litens, H. Hakansson, Y. Fery, D. Schrenk, M. van den Berg, A. H. Piersma and J. G. Vos, *Toxicology*, 2008, **245**, 109–122.
78. C. Zhou and C. Zhang, *Toxicol. In Vitro*, 2005, **19**, 665–673.
79. P. Ryden and R. R. Selvendran, in *Encyclopaedia of Food Science, Food Technology and Nutrition*, ed. R. Macrae, R. K. Robinson and M. J. Sadler, Academic Press, London, 1993, pp. 3582–3587.
80. M. Cheryan and J. Rackis, *Crit. Rev. Food Sci. Nutr.*, 1980, **13**, 297–335.
81. C. M. Weaver and S. Kannan, in *Food Phytates*, ed. N. R. Reddy and S. K. Sathe, CRC Press, Boca Raton, 2002, pp. 25–52.
82. S. Y. Choe, S. J. Kim, H. G. Kim, J. H. Lee, Y. Choi, H. Lee and Y. Kim, *Sci. Total Environ.*, 2003, **312**, 15–21.
83. M. B. Martin, R. Reiter, T. Pham, Y. R. Avellanet, J. Camara, M. Lahm, E. Pentecost, K. Pratap, B. A. Gilmore, S. Divekar, R. S. Dagata, J. L. Bull and A. Stoica, *Endocrinology*, 2003, **144**, 2425–2436.
84. I. Iavicoli, L. Fontana and A. Bergamaschi, *J. Toxicol. Environ. Health B*, 2009, **12**, 206–223.
85. X. Denier, M. E. Hill, J. Rotchell and C. Minier, *Toxicol. in Vitro*, 2009, **23**, 569–573.
86. L. Yuyan, W. Junqing, Y. Wei, Z. Weijin and G. Ersheng, *Fertil. Steril.*, 2007, **89**, 1008–1011.
87. R. J. Aitken and J. S. Clarkson, *J. Reprod. Fertil.*, 1987, **81**, 459–469.

88. M. Gavella, V. Lipovac, M. Vučić and V. Šverko, *Int. J. Androl.*, 1999, **22**, 266–274.

89. N. Sugino, Y. Nakamura, N. Okuno, M. Ishimatu, T. Teyama and H. Kato, *Biol. Reprod.*, 1993, **49**, 354–358.

90. M. Sawada and J. C. Carlson, *Endocrinology*, 1996, **137**, 1580–1584.

91. A. A. Alturfan, E. Zengin, N. Dariyerli, E. E. Alturfan, M. K. Gumustas, E. Aytac, M. Aslan, N. Balkis, A. Aksu, G. Yigit, E. Uslu and E. Kokoglu, *Folia Biol. (Prague)*, 2007, **53**, 183–188.

92. J. R. Arthur and G. J. Beckett, *Br. Med. Bull.*, 1999, **55**, 658–668.

93. P. Sidhu, M. L. Garg and D. K. Dhawan, *Drug Chem. Toxicol.*, 2005, **28**, 211–230.

94. A. Kumar, A. Malhotra, P. Nair, M. Garg and D. K. Dhawan, *J. Environ. Pathol. Toxicol. Oncol.*, 2010, **29**, 91–100.

95. P. Sidhu, M. L. Garg and D. K. Dhawan, *Chem. Biol. Interact.*, 2004, **150**, 199–209.

96. A. Goel, V. Dani and D. K. Dhawan, *Chem. Biol. Interact.*, 2005, **156**, 131–140.

97. N. A. Sadik, *J. Biochem. Mol. Toxicol.*, 2008, **22**, 345–353.

98. L. M. King, M. B. Anderson, S. C. Sikka and W. J. George, *Arch. Toxicol.*, 1998, **72**, 650–655.

99. H. Jemai, H. A. Lachkar, I. Messaoudi and A. Kerkeni, *J. Trace Elem. Med. Biol.*, 2010, **21**, 277–282.

100. D. Burukoğlu and C. Bayçu, *Bull. Environ. Contam. Toxicol.*, 2008, **81**, 521–524.

101. H. Rashid, F. Ahmad, S. Rahman, R. A. Ansari, K. Bhatia, M. Kaur, F. Islam and S. Raisuddin, *Toxicology*, 2009, **256**, 7–12.

102. J. D. Cook, M. B. Reddy, J. Burn, M. A. Juillerat and R. F. Hurrell, *Am. J. Clin. Nutr.*, 1997, **65**, 964–969.

103. L. Bohn, A. S. Meyer and S. K. Rasmussen, *J. Zhejiang Univ. Sci. B*, 2008, **9**, 165–191.

104. R. F. Hurrell, M. B. Reddy, M. A. Juillerat and J. D. Cook, *Am. J. Clin. Nutr.*, 2003, **77**, 1213–1219.

105. V. Ravindran, G. Ravindran and S. Sivalogan, *Food Chem.*, 1994, **50**, 133–136.

106. A. Viveros, C. Centeno, A. Brenes, R. Canales and A. Lozano, *J. Agric. Food Chem.*, 2000, **48**, 4009–4013.

107. K. Khanbabaee and T. van Ree, *Nat. Prod. Rep.*, 2001, **18**, 641–649.

108. D. K. Salunkhe, J. K. Chavan and S. S. Kadam, *Dietary Tannins: Consequences and Remedies*, CRC Press, Boca Raton, 1990, p. 177.

109. Proceedings of the 2nd North American Tannin Conference, Houghton, Michigan, June 17–21, 1991; *Basic Life Sci.*, 1992, **59**, 1–1053.

110. I. Barikmoa, F. Ouattarab and A. Oshauga, *J. Food Compos. Anal.*, 2007, **20**, 681–687.

111. FAO, *Food and Nutrition Series*, No. 27, Food and Agriculture Organization of the United Nations, Rome, 1995.

Subject Index

Note: page references in **bold** refer to figures, in *italics* to tables